LOCAL HERITAGE, GLOBAL CONTEXT

Amidst new talk of the "Big Society", we may forget that the concept of "place" has for long offered a multiplicity of small societies that are local and small scale yet universal to the human experience. This engaging collection of a dozen case studies presents a much-needed summary of the way that archaeologists today in Europe and beyond are thinking about place. For the authors, "sense of place" is firmly centred on people, and this book explores the multiple ways of deciding what this much-used term might mean, how sense(s) of place can be discovered, invented, promoted and used; ultimately, it makes us ask what is heritage actually for?

Graham Fairclough, English Heritage, UK

T0264649

Heritage, Culture and Identity

Series Editor: Brian Graham,
School of Environmental Sciences, University of Ulster, UK

Local Heritage, Global Context
Cultural Perspectives on Sense of Place

Edited by

JOHN SCHOFIELD
University of York, UK

ROSY SZYMANSKI
English Heritage, UK

Routledge
Taylor & Francis Group

LONDON AND NEW YORK

First published 2011 by Ashgate Publishing

Published 2016 by Routledge
2 Park Square, Milton Park, Abingdon, Oxfordshire OX14 4RN
711 Third Avenue, New York, NY 10017, USA

First issued in paperback 2016

Routledge is an imprint of the Taylor & Francis Group, an informa business

British Library Cataloguing in Publication Data
Local heritage, global context : cultural perspectives on
   sense of place. -- (Heritage, culture and identity)
   1. Place attachment--Congresses. 2. Community-based
   conservation--Congresses. 3. Cultural property--
   Protection--Congresses. 4. Community-based conservation--
   Europe--Case studies--Congresses. 5. Community-based
   conservation--Australia--Case studies--Congresses.
   6. Cultural property--Protection--Europe--Case studies--
   Congresses. 7. Cultural property--Protection--Australia--
   Case studies--Congresses.
   I. Series II. Schofield, John. III. Szymanski, Rosy.
   363.6'9-dc22

Library of Congress Cataloging-in-Publication Data
Schofield, John.
   Local heritage, global context : cultural perspectives on sense of place / by John Schofield
and Rosy Szymanski.
      p. cm. -- (Heritage, culture and identity)
   Includes index.
   ISBN 978-0-7546-7829-8 (hardback) 1. Cultural property. 2. Place attachment.
   3. Environmental psychology. I. Szymanski, Rosy. II. Title.
   CC135.S324 2010
   155.9--dc22

                                                                    2010029864
ISBN 13: 978-1-138-24833-5 (pbk)
ISBN 13: 978-0-7546-7829-8 (hbk)

List of Contents

List of Figures

List of Maps

List of Tables

List of Contributors

Anne Brakman	Municipality of Maastricht, The Netherlands
Duncan Brown	English Heritage, UK
Sue Clifford	Common Ground, UK
Antony Firth	Wessex Archaeology, UK
Rodney Harrison	Faculty of Arts, Open University, UK
Cathy Hopley	Forest of Bowland AONB, UK
Paul Mahoney	Countryscape, UK
Hilary Orange	Institute of Archaeology, University College London, UK
Rachel Radmilli	Department of Management, University of Malta
John Schofield	Department of Archaeology, University of York, UK
Rosy Szymanski	English Heritage, UK
Stephen Townend	Entec UK Ltd, UK
Paula Uribe	University of Saragossa, Department of Antiquity Studies, Spain
Gurly Vedru	Institute of History, Tallinn University and National Heritage Board, Estonia
Ken Whittaker	Entec UK Ltd, UK
Jason Wood	Heritage Consultancy Services, UK

Preface

This volume has an interesting history. It began with a big question about the significance of local places in what is often a predominantly national or at best regional heritage discourse, countering the scant regard often paid to local views, and promulgating the idea of NIMBYs (those arguing against development or change 'in my back yard') as mere busy-bodies. We believe that while these broader frameworks are necessary and perform a useful (indeed essential) role, it is local communities who are often the real experts, who are 'experts at living where they do' and who best understand the impact of change on their local environment. That was the question, which one of us (Rosy Szymanski) then took up during a placement with John Schofield at English Heritage during her studies for an MSc in Professional Archaeology at the University of Oxford's Department of Continuing Education, resulting in a database of definitions and interpretations of sense of place gathered from a diverse range of organizations and individuals.

A second stage, determined by the conclusion of this study, involved organizing two related conference sessions, first at the Institute of Field Archaeologists (IFA, now IfA – Institute for Archaeologists) conference in Swansea (2007), followed by another at the European Association of Archaeologists conference in Malta (2008). All of the chapters in this collection were presented at either one of these conferences, the calls for papers for which were near identical, noting how 'sense of place',

> is such a familiar phrase, and one now commonly used in professional and domestic situations to describe the emotional attachment people have to the places they hold dear. This sense of place – sometimes referred to as '*genius loci*' – can equate with what has been termed 'the lure of the local', with its concern for the familiar – the place where we live, or where we lived when we were children. It is also about rootedness, belonging, stability and identity. But that is a very broad definition and perhaps not so helpful in policy work, and developing community-based heritage projects that seek to assess or characterise local areas. As heritage management practices increasingly take account of 'the local', and draw on the views and expressions of interest amongst local communities (people telling *us* what matters to *them*), the need to fully understand what is meant by sense of place, and its uses and implications, becomes arguably more important than just semantics.

This book, like the conference sessions that preceded it, comprises a series of contributions that together review what is meant by sense of place, and why (or whether) it is useful in the context of heritage management practice. It examines how we can define sense of place, what precisely it means, and what its implications are for managing change now and in the future. We explore these contextual issues further in our introduction.

We would like to take this opportunity to thank all of the contributors to both conference sessions, including those who chose not to participate in the publication and those notable others who contributed so helpfully to the discussions in Swansea and Malta. We are grateful also to our employer English Heritage for enabling our attendance at these events, and the conference organizers at IfA and EAA for accepting our proposals. We extend our thanks also to staff at Ashgate, notably Sarah Horsley, Valerie Rose and Caroline Spender, for easing us through the production process. Finally, we would like to thank all of our contributors for their enthusiasm and commitment to seeing this project through to completion. We hope the final product measures up to expectations.

John Schofield
Rosy Szymanski

Chapter 1

Sense of Place in a Changing World

John Schofield and Rosy Szymanski

What are we to understand by the term 'sense of place', and how can it be helpful in the context of cultural heritage practices? How can we measure or capture sense of place, and should we even try? And what happens when strongly-held and personal views come into conflict, either with each other or with those of authority? Our sense of place project, culminating now in this edited collection of ideas, examples, conversations and suggestions, will explore these and other questions through a series of applications in which sense of place is a central concern. In this introduction we describe some of the underlying principles, setting the scene for the more detailed contributions that follow.

Change

The world is changing, as the world has always changed, and as archaeologists and heritage practitioners this is something we understand only too well. We know for example that people have always reacted to change and responded to it in different ways, welcoming it or being wary and suspicious of what it might entail, fearing the unknown. Of course, change may be expected and lengthy preparations can be made for what is to come. But change can also be unexpected and traumatic. It can take us by surprise. Change happens in different ways, at different scales and at different speeds. It can be immediate or it can be gradual. Sometimes we hardly notice that it is happening – the creep of progress, the cumulative effect of which can transform and reconfigure to the point where something becomes fundamentally different, though it can be hard to establish when 'the change' occurred.

As archaeologists we often record and witness through our surveys and excavations the process of change over longer periods. But our dilemma as archaeologists operating within the heritage sector is that we can also shape the world through our heritage practices. Replacement doors and windows on properties in a protected conservation area for instance: replacing one front door may not detract from an area's character; removing two even three may not make much difference. But at some stage the degree of change does have a detrimental affect and the prevailing characteristics once sufficient to justify the conservation area are lost.

A particular concern is the potential impact of changes which have *social* significance, in the sense that they affect people's lives, or the quality of their lives. How might change affect or compromise a community's, or a person's sense of place? Such changes can obviously be physical (an inappropriate new building close to or instead of an older one), but they can have social impact as well, as Grenville has highlighted in her essay on ontological security (Grenville 2007). Take the example of an architecturally mundane but socially meaningful place (even a type of place) within a community, a youth club maybe, or an old cinema, any place that holds memories for members of the community and is valued as a result. If planners suggested removal of that place, to accommodate more housing for example, the community's response might be critical and perhaps hostile. Local heritage practitioners might share the community's concerns and take their side, or they might side with the planners in some cases, where they felt local opinion was misguided or where adequate mitigation was seemingly in place. The simple point here is that local perspectives do matter, and are often grounded in a strong emotional connection to the place that is threatened. Some dismiss these local concerns as merely the views of NIMBY's (Not In My Back Yard), local busy-bodies and people who merely stand in the way of progress. But as Burström et al. (2004) and others have said, this calls into question who precisely the experts are; is it the planners and heritage practitioners, qualified to take rational, objective views on the basis of experience and regional or national context, or is it the local people, who know the place and its capacity for change best of all, people who are, after all, experts at living where they do?

This book, ultimately, is about change. But it is also about recognizing, documenting, understanding and taking account of what local people value about their local environment, and the processes by, and degree to which these 'special' things can be retained amidst the inevitability of a fast-changing world.

Local

To introduce a short discussion on local-ness, we want first to consider this word 'special'. By special we do not necessarily mean iconic. We are not in the same territory here as words and values that are specifically used to justify the introduction of heritage protection measures, such as listing buildings of 'special historic interest' and monuments of 'national importance'. Rather, we are typically referring here to things (which can mean places, objects, cultural traditions, landscape components) that are valued locally, that characterize a local area, that give a place distinctive quality, that set it apart from other places. Of course, some of these places are 'special' in terms of cultural significance, national importance and so on. But more often they are not. They are ordinary, mundane, everyday places, the commonplace in national terms, but deeply ingrained with local significance and special to those who live there. Such special things need not always be tangible. As Tuan explains, 'odours can lend character to objects and places, making them distinctive, easier to identify and remember' (2005: 11). Sound can also be distinctive and can evoke spatial impressions. Musical traditions can be highly localized, while a place's auditory characteristics can offer distinctive qualities. Taste can also have a close proximity to place, in local culinary traditions for example. We are talking here about things that contribute to sense of place, or – in Tuan's words, '*genius loci*'. Those things need not be (and often are not) material.

All of these contribute to local character. And local is important, especially now given the 'ultimate abstraction of reference that derives from the post-modern condition that has the author him- or herself as the object of study', as Tom Conley notes in his introduction to Augé's *In the Metro* (2002: xvii, and cited in Harrison and Schofield 2010: 135). For Augé, 'solitude accrues as the world accelerates' and there seems little doubt the world is accelerating (cf. Glieck 1999, but cf. Edgerton 2006 for an alternative view).

The world is also one in which migration (forced and selected), diaspora and transience are now commonplace. At least in the developed world, and increasingly (for all the wrong reasons) elsewhere, people are on the move. People migrate and commute over ever-increasing distances, thus ensuring a loss of contact with the places that mean most to them, and the places they call home. Of course movement means making new connections with new places, and learning how to inhabit an unfamiliar landscape with unfamiliar mores and cultural traditions. Yet these connections can also be close, even though they are different in form and intensity to those that exist for a home or homeland. For example, the Faro Convention (Council of Europe 2009) has relevance here, with its stated aims of putting 'people and human values at the centre of an enlarged and cross-disciplinary concept of cultural heritage' and 'recognizing that every person has a right to engage with the cultural heritage of their choice, whilst respecting the rights and freedoms of others, as an aspect of the right freely to participate in cultural life.'

While 'Faro' almost certainly marks an important threshold, it is important to recognize that this whole area is fraught with tensions and difficulties. This is because, ultimately, sense of place is a personal matter; it is what individuals often think matters most, and what it is that characterizes a neighbourhood. It is something people feel strongly about. As Peter Read has explained (1996: 3),

> People respond individually to locality ... and the culture with which they are familiar helps to enlarge, diminish, shape or transform it. Senses of belonging are allied to attachment and love, but the country must first become known and apprehended.

Knowing a place or landscape is relative. Knowledge can be accumulated over generations, or over weeks, days even. These familiar places and areas of landscape are also reference points which, according to Relph (1985), construct in our memories and affections, a *here* from which to discover the world, and a *there* to which we can return. Naturally enough, therefore, opinion on the value of these places and areas will vary, often significantly, and often these differences of opinion will clash, at places that some value highly and some detest or find troublesome. In the case of a rural community with a large migrant population for example, those that have lived there for generations will inevitably feel a sense of ownership of 'their' place, knowing it more intimately, and having memories and stories woven into its fabric. As a result, they will probably feel their view should prevail. But that is to deny recent migrants a say. They may not have lived there for as long, but it has nevertheless also become their place, and they too will have views and opinions about it. Those views will be shaped in part by the landscape and cultural traditions prevalent in the landscape they have come from, views (and their attendant practices) which may be at odds with those of the existing community.

This then introduces a further set of issues. How should these tensions be addressed with a view to achieving balance and resolution; and how might practitioners wanting to gather information about sense of place negotiate such complex and challenging ideas with a mixed community, and where language and cultural barriers might exist? There is also the further complication, identified by Tuan (2005: 136–7), that:

> Intimate experiences lie buried in our innermost being so that not only do we lack the words to give them form but often we are not even aware of them. When, for some reason, they flash to the surface of our consciousness they evince a poignancy that the more deliberative acts – the actively sought experiences – cannot match. Intimate experiences are hard to express.

Hard, maybe, but not impossible. As Tuan goes on to say, these intimate experiences may be personal and deeply felt but they are not necessarily solipsistic or eccentric. Hearth, shelter, home or home-base are intimate places to human beings everywhere, and often form the basis for evaluating 'sense of place'. Home is the ultimate in local. Here we briefly assess these two questions under the headings 'Engagement' and 'Home', using the opportunity also to introduce the chapters that follow, and which negotiate these areas of tension and complexity with greater focus and clarity.

Engagement

A recent survey in England (Bradley et al. 2009) emphasized the importance of the historic environment as contributing to sense of place. But equally if not more important, it demonstrated the extent to which people in the UK first, understand their local environment, and second, take opportunities for engagement with it, with alacrity and enthusiasm.

But as we have seen, this is an area often fraught with tensions and difficulties. What happens when views and perspectives on a place are at odds, or even in conflict; where valued places for one set of people are loathed by another? And more straightforwardly, what methods exist for documenting and negotiating these values? Martin Thomas' study of the Macedonian community in Sydney is an example of how this can be resolved, with an approach that is inclusive, respectful and rigorous. To set the scene, he introduces Paul Stephen, a Macedonian-born Australian born in a mountainous region of Macedonia in 1936, before emigrating to Australia in 1948. Paul described the landscape of his childhood:

> Martin, you're gonna make me cry now. I'm here because of that landscape. We have the most wonderful landscape. I don't remember drought. I don't. We were fairly north; we would have been about 800 to 1,000 metres above sea level. We had plains, and on our plains in fact was originally an ancient city there. We could see these beautiful mountains where the forest was from the village. It always had snow on the peaks. But he (my uncle) always said to me there's a lake there and this is where heaven is. And not until 1984 was I allowed to enter my area, and to one of my cousins I said: 'Look, I have to go to this lake.' And my uncle was right, it is heaven. Because there's no tourists there. No pollution. The white ducks are still there, the black ducks are still there. I cried with happiness. It is so supreme, so silent, so beautiful, and you've got these big pines and they're huge. (Thomas 2001: 7)

This attachment to a hilly, wooded landscape leads the author to a discussion of the huge Macedonian picnics that occur annually in Royal National Park, Sydney, involving large numbers of Macedonian migrants. It was the woodland, and the familiarity of woodland, that drew them to this place. The events are seen by others as 'rowdy, congested and environmentally unfriendly' (ibid. 8). Yet for Macedonians they represent an important social tradition held in a familiar landscape. Paul explained how the large picnics, where people could eat, drink and play music, were a hybrid tradition, influenced by outdoor celebrations that occurred in the homeland though inflected by the Australian context. Thomas' (2001) study identifies the tensions, before presenting a methodology for achieving some resolution, through mutual understanding. The methodology concerns approaching and engaging the various communities and negotiating and discussing the issues of use and management with them. Here two sets of values, both deeply ingrained with a sense of place, are balanced in a way that allows each community of interests to be recognized and respected.

Within cultural heritage practice some procedures and methodologies can be unduly prescriptive and mechanistic, glossing over the complexities of subjectivity and professional judgment for example. But sometimes the methods need to accommodate and embrace the complexity being addressed. As Thomas' example demonstrates, this can be achieved. In this collection of essays several examples are presented. Sue Clifford's chapter, for example, presents the innovative, ground-breaking and acclaimed work of Common Ground, demonstrating through numerous examples and approaches how community participation can be both fun and empowering. The Parish Maps for example demonstrate with extraordinary eloquence and skill, the places and things valued by a local community. These are not prescriptive or exclusive in any sense. The maps are consensual and inclusive of all who choose to be involved. In Cornwall, Hilary Orange uses questionnaire surveys to closely examine sense of place, and how perception and experience is influenced by issues surrounding Cornish-ness. A particular emphasis here is the tension that exists between a very beautiful and predominantly coastal landscape, and the traces of ugly, dirty, noisy industry, albeit now monumentalized and incorporated into the aesthetic. Asking questions and analysing responses moves us closer to understanding Cornish-ness and sense of place amongst participants from Cornwall and elsewhere, and amongst people of different gender- and age-groups.

A more obviously geographic focus is presented by Rodney Harrison who follows a general discussion of sense of place within the context of Australian heritage practice with a review of methods such as mapping attachment and counter-mapping (generating maps that challenge order and authority). Using examples from New South Wales and also south London (UK) he describes how simple cartographic and ethno-historic practices can contribute significantly to mapping attachment. Influenced by Australian heritage practice and examples such as local community capacity building in South Wales, Ken Whittaker and Stephen Townend also focus on methodology, describing the development of Qualitative Data Analysis. This is a technique they have adopted in their role as consultants working in the UK planning system, to assess and draw out understandings of the value of the historic environment. Their chapter outlines the concern that this is a relatively new area for heritage practitioners who, until fairly recently, have focussed their attention on the physical remains of the historic environment rather than trying to tease out and negotiate the values associated with such places.

As well as discussing methods for understanding the attachments and values associated with existing, known and familiar places, the authors in this book also discuss the question of shaping the connections a community has to a place and perhaps helping them to realize a connection with the remote or unfamiliar. Anne Brakman, an archaeologist working in Maastricht, the Netherlands, was challenged by a developer to show how cultural heritage could 'enforce the quality of life' in a newly constructed settlement and attract potential home buyers. The association between quality of life and rootedness is something often discussed by government organizations and local planners but it is not easy to apply in the design of a new neighbourhood. Anne Brakman looks at ways to build connections between the past and new inhabitants and to encourage them to think of new development (and their life in that new neighbourhood) as just the latest chapter in an unfolding story – the latest (not the last) of many layers, in archaeological terms. Antony Firth looks at connections with what for many is an unfamiliar environment, the sea. He asks whether the sea, which lacks fixed surroundings and is in constant motion, can generate a sense of place. After reviewing theoretical understandings of perception and how individuals come to understand and perceive places, he looks at the sense of place we may experience from wrecks of ships and aircraft and from submerged areas which were once dry land. He argues that our sense of place from the sea is no less authentic for being experienced remotely or indirectly through electronic means. He also emphasizes the importance of archaeologists learning to share their interpretations and understandings of place with the wider public who fund this work.

Artistic practice connects people to place in imaginative and often unforeseen ways. Sue Clifford discusses this in her chapter, using Parish Maps for instance, Rachel Radmilli describes the value of documentary film-making while Jason Wood's example from Middlesbrough demonstrates how artistic intervention can mitigate the sense of trauma so keenly felt when one's home (or in this case a much-loved place that felt like home – a football stadium) is lost. Again there is an eloquence here often absent from heritage discourse. But why is that? Heritage is a personal thing as well as (if not more than) being cultural and corporate. That fact is clearly evident in Jason Wood's chapter, including through his illustrations and his choice of an opening quotation from Nick Hornby, highlighting the novelist's obvious strong attachment to a particular part of north London.

As we have said, this collection is not definitive and one area lacking is the role of new communicative technologies. This is a fast emerging field and was in its infancy as this publication took shape. Nevertheless as a truly inclusive facility, new technologies – and specifically we refer here to GPS enabled mobile phones – have the capacity to connect everyone in society to the democratic processes of place-making and planning for the future.

What draws all of these methodologies together is that they are: enabling (in that they get things done); they are empowering and participatory; they do not discriminate; and they are – more often than not – fun to engage. In 2008 the artist Christian Nold coordinated a project in west London called the Brentford Biopsy, which saw locally-held values and perceptions inscribed on a physical map of the area ultimately exhibited in a local gallery space (see also Nold 2009). By all accounts this was fun, constructive and inclusive. Christian Nold describes its purpose, and its ambitions:

> This map is for anyone who cares about Brentford or who really ought to care about it. The hope is that all the Brentford stake-holders such as local people, interest groups, developers and the council will use this map to revitalise their discussions. In particular, we insist on the role of people's sensory and emotional experiences as an essential part of all political discussions. How each one of us 'feels' about each other and our environment is the foundation stone upon which any democratic decision-making has to be based. To do this, we first have to enable people to focus more strongly on their own experiences, reflect and question them and then to articulate and share them through a political process where their personal experiences are valued and not disregarded. The challenge that this map presents to all the local stake-holders is how to use this document productively and include it within the process of politics. It is this complexity and difficulty of how to situate the content of this map that should also give it a value and meaning to a wider audience who are interested in new ways to represent local and intra-local issues or ways to instigate local public spheres. (http://www.publicbiopsy.net/info.htm – accessed 20 April 2010)

Such is the challenge facing all of the ideas and initiatives presented here.

Home

As Peter Read has said (1996: 101), homes – like other places – are mentally constructed. They are obviously often physical places as well, but the mentally constructed home is arguably the more significant. It is also uniquely 'ours'. One person's home is different to another's, even though they might overlap. In one example, Read describes how home can be a large town or city, a suburb or area within the city, a rural community, a house, a room in a house, even a tree in the garden. Home, as T.S. Eliot said, 'is where one starts from'. Home is the root of most conceptions of sense of place. Paula Uribe looks at ways in which members of Roman society attempted to overcome homesickness by recreating some of the familiar surroundings of their native places in the new environments of the territories they conquered. Her study focuses on the Iberian Peninsula, looking at domestic architecture as the most intimate expression of sense of place. Roman customs, practices and architectural styles then became part of a process of cultural exchange adopted and adapted by the indigenous population.

In Gurly Vedru's chapter on values and place in Estonia, some interesting and rather surprising opinions are expressed and analysed, following a questionnaire survey. Home here appears to be a slightly confusing concept, or at least one not deeply rooted in the past. It is comparatively unusual to hear of economic value outweighing cultural values and sense of place, but that does appear to be the case in Rebala where the majority of the population is indigenous. You would expect a strong connection to the landscape. But as Vedru concludes, most members of the local community want to change their surroundings, sell their land in parcels and build new houses. The landscape in itself does not have high emotional value for the local community. It is more a landscape laden with potential economic value. In the area of Tõdva-Kajamaa-Lokuti the response was very different. Here a population of recent arrivals is more concerned to preserve sites that reflect the area's historic character. Here a rootedness is evident where one would not necessarily expect to find it.

Home is also the subject of Rachel Radmilli's essay on Valletta, which describes the film *Ilhna Beltin*, commissioned as part of a wider Mediterranean project on community and local identity in various Mediterranean cities. Valletta may be a changing place, but it is home to the project's participants. One woman describes in the film how she left Valletta once to emigrate to Australia, but would never leave the city again. In fact she didn't. She died recently keeping her promise, remaining loyal to her city of birth. The film and the chapter that describes it also present the story of a recent migrant (the film-maker), who had difficulty settling in Valletta but for whom the process of film-making and documentation strengthened his own sense of belonging.

Home for Jason Wood is not a domestic space so much as a place of regular and routine pilgrimage: the football stadium. His chapter describes eloquently and persuasively how these places accumulate social significance, a value that remains closely woven into the fabric of the stadium and the family lives of the team's supporters. Equally eloquent are the expressions and manifestations of loss: the trauma that accompanies change and closure, and the need to hold on to fragments of what has gone.

Home was the starting point for developing the 'Sense of Place Toolkit' for the Forest of Bowland Area of Outstanding Natural Beauty, the subject of the chapter by Cathy Hopley and Paul Mahony. The aim of the toolkit was to raise awareness of the Forest of Bowland externally and to encourage local people and businesses to develop an affinity and loyalty to the area. Prior to the creation of the toolkit the Forest of Bowland was often promoted by local people as a useful stopping point between better known destinations like the North Yorkshire Moors or the Lake District. Inspired by the work of organizations like Common Ground, the creation of the toolkit drew heavily on the stories and views of those living and working in the area in order to draw out the special qualities of that particular place. The authors feel this approach has resulted in the creation of a brand, a sense of place, which truly belongs to the area and its inhabitants.

Commercial interest, in the sense of marketing and branding sense of place, is also the starting point for Duncan Brown's chapter on archaeology and identity in his home town, Southampton. He looks at how much the study of the past, and archaeology, actually informs a sense of identity in a particular place and how archaeological interpretations enter the public consciousness. The trauma that followed the loss and destruction of Southampton during the Second World War made the post-war community keen to re-establish its roots with the city's past. The archaeological discoveries made during the rebuilding of the city, the creation of new city museums and the restoration of its monuments were then a source of local pride. But today it seems there is a lack of integration between the value the inhabitants of the city put on the richness of their archaeological past and the way this is used to inform planning for the future of the city. Planners and local politicians prefer instead to focus on those aspects of the city's past that lend themselves to the creation of 'themes' and advertising slogans.

Conclusion

Ultimately, as John Agnew (1987) said, sense of place is the subjective and emotional attachment people have to place. It conforms closely to what Lucy Lippard (1987) refers to as, 'the lure of the local'. According to Tim Cresswell (2006: 8), it is commonplace at least in Western societies in the twenty-first century to bemoan the loss of sense of place, 'as the forces of globalisation have eroded local cultures and produced homogenised global spaces'. This book takes this suggestion as a starting point, proceeding to explore means and mechanisms by which sense of place can be understood, assessed and accommodated in the increasingly democratic process of managing change, at least as it is represented through heritage practices in the developed world. The examples are intended not to be definitive, and nor necessarily to be representative of an emerging, fascinating and challenging field of research. They are however statements on what can be achieved and the methods and processes by which sense of place can be taken account of, successfully balancing heritage interests with progress and change, all in the name of sustainability.

References

Agnew, J. 1987. *The United States in the World Economy*. Cambridge: Cambridge University Press.

Augé, M. 2002. *In the Metro*. Minneapolis and London: University of Minnesota Press.

Bradley, D., Bradley, J., Coombes, M. and Tranos, E. 2009. *Sense of Place and Social Capital and the Historic Built Environment*. Unpublished Report of Research for English Heritage.

Burström, M., Elfström, B. and Johansen, B. 2004. Serving the Public: Ethics in Heritage Management. In H. Karlsoon (ed.), *Swedish Archaeologists on Ethics*, 135–47. Bricoleur Press.

Council of Europe 2009. *Heritage and Beyond*. Strasbourg: Council of Europe Publishing.

Cresswell, T. 2006 (2004). *Place: A Short Introduction*. Oxford: Blackwell Publishing.

Edgerton, D. 2006. *The Shock of the Old: Technology and Global History since 1900*. London: Profile Books Ltd.

Glieck, J. 1999. *Faster. The Acceleration of Just About Everything*. London: Abacus.

Grenville, J. 2007. Conservation as Psychology: Ontological Security and the Built Environment. *International Journal of Heritage Studies* 13.6, 447–61.

Harrison, R. and Schofield, J. 2010. *After Modernity: Archaeological Approaches to the Contemporary Past*. Oxford: Oxford University Press.

Lippard, L. 1997. *The Lure of the Local: Senses of Place in a Multicultural Society*. New York: The New Press.

Nold, C. (ed.) 2009. *Emotional Cartographies – Technologies of the Self*. Available from www.emotionalcartographies.net.

Read, P. 1996. *Returning to Nothing: The Meaning of Lost Places*. Cambridge: Cambridge University Press.

Relph, E. 1985. Geographical Experiences and Being-in-the-World: The Phenomenological Origins of Geography. In D. Seamon and R. Mugerauer (eds), *Dwelling, Place and Environment*, 15–31. Dordrecht: Martinus Nijhoff.

Thomas, M. 2001. *A Multicultural Landscape: National Parks and the Macedonian Experience.* Sydney: NSW National Parks and Wildlife Service and Pluto Press, Australia.

Tuan, Y.-F. 2005. (1977). *Space and Place: The Perspective of Experience.* Minneapolis and London: University of Minnesota Press.

Chapter 2
Local Distinctiveness:
Everyday Places and How to Find Them

Sue Clifford

A place holds more than any guidebook, novel or academic treatise can tell you – for it implies many dynamic relationships between people and geography. The differences between places are amplified by time and the sedimentation of memory. Everywhere is somewhere to someone – the land, embossed by story on history on natural history, carries meaning. It is through meaning that attachment, watchfulness and rapport are forged. Our future places need culture and nature to intertwine so well that there is room for both and richness in each.

Context

Common Ground began making arguments for the local and the commonplace in 1982/3. It seemed to us that the task of informing and awakening ecological conscientiousness was vital – the professionals would always be too few and their expertise needed the grounding of local knowledge and altogether more voices. Common Ground's '*grand projet*' needed the reach, the knowledge, the power of people everywhere. Yet the scale had to be local, where knowledge and familiarity might build upon and breed attachment and action.

It was hard going to even achieve a hearing from experts that everyday nature, unwritten history, and vernacular buildings were significant and in danger; that the special would never survive without the ordinary, that everyday surroundings and their inhabitants were important anyway.

We were striving to articulate the importance of everyday surroundings to local people in a professional world driven by concern to protect the rare, the endangered, the spectacular. In their often prim, segregated pigeonholes of nature, history, archaeology, landscape, countryside, buildings, and ancient monuments, professionals communicated across boundaries with difficulty. Local knowledge was rarely sought or admitted, wisdom hardly remembered as an aim. In the 1980s loss and deterioration were increasing – wild life (King 1980), land and landscape (Shoard 1980), old buildings, ancient monuments (Baker 1983). Alienation was reportedly multiplying. Homogenization of high streets, fields, and front doors was already under way.

By the second decade of the twenty-first century, loss continues to escalate, protective legal frameworks have been eroded, statutory bodies have been pruned and amalgamated, participation and the local have been recognized as important but hard-pressed professionals find less and less time to achieve what should take a long while. Powerful interests press development through; economics still dominates.

And climate change, whilst terrifying, is merely a symptom of our estrangement from where we live. The real malaise lies in our relations with Nature.

Contentions

Common Ground has always contended that we need to re-value our relations with Nature in the deepest philosophical sense, as well as very practically in our everyday surroundings. And we need the engagement and action of everyone.

We have to re-educate ourselves to live together and to live with nature. The web of understandings between people and the land, people and their histories, is not about scenery, it takes us below the surface, to where the land might reflect back to us purpose and belonging.

For Common Ground, driven by the ecological imperative, focusing on 'place' gave us a way into interconnections and helped us to ground them. The mix of objective and subjective which gives a place significance comes from the plurality of interactions that have helped people to understand and shape the land, from cities to fields.

It is not the castle, cathedral or cuckoo pound, nor even burial mound, bus-shelter or bluebell wood that separately defines the significance of a place, but a messy mingling of things tangible and intangible, fixed and transient, big and small, ordinary and special.

Shared understanding between professionals and locals would stop the building on areas that sometimes flood, would reveal where children used to catch great crested newts, would tell the stories of ancient mining, crashed war planes, old wells. If democracy is to have a real future it demands a fractioned way of seeing and understanding, helping wise decisions all the time, not simply voting once in a while, or not.

For Common Ground significance is not about the special or the separated out. Life is lived in 'places' by people; they invest places with meaning. Professionals and politicians have seemed blind to both.

Entangled in all of this is the spectre of the receding democratic voice. Careful decisions about places should take a long time, should involve as many as possible; place and decision-making become their own academy of democracy – we all learn. In the framing of policy we feel it vital that as much knowledge and as many people exchange ideas and knowledge to encourage and inform the best everyday actions – tumbleweed expertise meeting indigenous knowledge should give us better decisions. The need is for a more mature, involving and fractioned democracy, a nurturing of pluralism.

So much surveying, fact gathering, desk research, analysis and policy-making leaves out the very things, the nuances, that make a place significant to people who know it well. And if all you do is try to protect the rare, the special, the spectacular and you do it by putting a red line around it, you condemn the rest. You risk undermining and alienating the very allies you need most – the local people. The familiarity that builds attachment and identity is a two way street – people feel and take responsibility if there is a way in.

The projects that Common Ground has initiated have offered ideas to enthuse people into questing for more knowledge about their place and then into action based upon their findings. For us it has been important to lure people into adventuring together, demonstrating to themselves how much they know, and to discover how exciting it is to find out more by looking, delving and asking. We have tried to devise ways of encouraging people to simply get started, to observe, record and share information and values about their place. The hope is that they discover that working together is enjoyable, that people have an extraordinary range of knowledge, that things normally marginalized really are valued, but perhaps embarrassment at caring for something so workaday stopped people saying so. In parallel we have tried to engage the professionals in welcoming local involvement and being imaginative in exchanging information with people.

For us, importance lies in the rapport and shifting edges between nature and culture. This rich mix is often uncountable, unquantifiable, but far from being the 'soft' subject matter professionals have been guilty of underplaying or even scorning, the intangible, is still being marginalized because it is just too hard to deal with.

When we put together the words *Local Distinctiveness* we were trying to animate places in local minds and lives, to help people descriptively engage with each other about their place – to articulate meaning. Not about specialness, but about everything – assemblage and accumulation: difference, detail, patina, authenticity, identity – *Local Distinctiveness* has been our mantra to help people discuss their sense of place since our first book 'Holding Your Ground' (King and Clifford 1985). This has been achieved for example through our 'Parish Maps' initiative.

Parish Maps

One of our earliest projects from the 1980s still rolls on, and many Parish Maps have been at the root of community action in town and country.

'Parish' is used to suggest a small place defined from within, 'Map' implies the capture of four dimensions into two or perhaps three, all to be worked out by people in the community. Some choose to work with an artist/designer/director who helps draw the pieces together into a coherent entity – a printed poster, a giant tapestry, a photo collage, a back cloth for a community play, a set of flags, a film.

We have found that a subjective question gives people a way in (starting is often the most difficult task). 'What makes this place different from the next' and 'What do you value here?' makes them the expert, builds confidence, gives them a chance to express what is of significance to them. No-one can tell them what they care about. Suddenly the community begins to share stories, old and new, observations made over time, materials and artifacts long thought vanished, names pronounced in the local manner that explain the underlying meaning, memories of how things have been moved or obscured by nature. Academic knowledge gets added, libraries are visited, land is walked, photographs are taken, things are counted, professionals are asked for advice, celebrations emerge or are revived which in turn uncover new participants and knowledge.

It takes time to reach beyond the stereotype, so many cursory projects reduce place to postcard or wring it out in abstractions. But given time people, having told each other the same story and pointed to the expected landmarks, break through into the things long taken for granted, half forgotten or discretely known.

There is also a need to liberate people from the, albeit wonderful, formality of the Ordnance Survey map. The OS map is of course full of subjectivity, mistaken spellings, old fashioned pictures of conifers and deciduous trees as well as new fangled contours: it masquerades as objective, true to scale. It is not infallible, it may not tell the truth. We wouldn't be without it, it has evolved into a beautiful and useful thing, but it is not entirely accurate or the only way to communicate by map.

The Thirsk Parish Map is 24 feet long and 6 feet high. It took two years to research and nine months to complete. The project began when this Yorkshire Civic Society became concerned that 'things can all too easily disappear before people realize what is happening ...'. The map now hangs in the library as a permanent record with a powerful influence on future development. They felt that 'if outside developers had some idea before they make plans for an area just how the community feels, they would think twice before investing their money in destroying things that other people value.'

Hanging the Parish Map in a public place ensures that conversations continue.

A Few Examples

Figure 2.1   Detail of Muchelney Parish map
Source: © Gordon Young 1987

An Artist's Map   In Muchelney (Somerset) visual artist Gordon Young (1987) created his own painted map in black and white (Figure 2.1). It deftly exposes wet moor, a place of toad spawning, orchards, window and door details of certain houses, abbey relics, Geoff's hedge (holly, clipped to exacting angles), and lightning striking a holm oak tree, all amongst the layout at the crossing of roads and with more and more detail to explore. An angel brandishing 'Peace on Earth' seems ironically unruffled by a low flying fighter-plane from nearby RNAS Yeovilton, an eye and arrow point to a singular hill with a tree on top, an eel, a dragon and two tractors share the fields with pollarded willows, heron, kingfisher, cows and mushrooms. The push and pull of history jumbled and active in legend and hard work brings dynamism to the topography. Some corners are marked 'unexplored'. It is personal, and communicates with passion and relish the rag-bag of things that for him give meaning to this place.

A Community Map   In Derbyshire, the people of Bonsall researched, gathered and sifted the minutiae of their old mining community (Figure 2.2). Into their printed map a wealth of information is laid out in a bird's eye view of the village and its environs. Holding it all together is a border monologue, hand copied, that chills the soul. It places one not merely into a landscape, but into an economy, a social system, a particular weaving of history related to the land:

> BY CUSTOM OLDE IN WIRKSWORTH WAPONTAKE
> IF ANY OF THIS NATION FIND A RAKE
> OR SIGN LEADING TO THE SAME; MAY SET
> IN ANY GROUND & THERE LEAD ORE MAY GET
> THEY MAY MAKE CROSSES, HOLES OR SET THEIR STOWES
> SINK SHAFTS, BUILD LODGES COTTAGES OR COES
> BUT CHURCHES, HOUSES, GARDENS ALL ARE FREE
> FROM THIS STRANGE CUSTOM OF THE MINERY
> FOR STEALING OAR TWICE FROM THE MINERY
> THE THIEF THAT'S TAKEN FINED TWICE SHALL BE
> BUT THE THIRD TIME, THAT HE COMMITS SUCH THEFT
> SHALL HAVE A KNIFE STUCK THROUGH HIS HAND TO TH' HAFT
> INTO THE STOW, & THERE TILL DEATH SHALL STAND
> OR LOOSE HIM-SELF BY CUTTING LOOSE HIS HAND
> & SHALL FORESWEAR THE FRANCHISE OF THE MINE
> AND ALWAYS LOSE HIS FREEDOM FROM THAT TIME

Views across fields broken with drystone walls, rubble and grassed heaps, field barns and coes (little stone 'barns' topping out the old mine shafts) reveal the worked topography. Portraits of the parasol mushroom, cowslip, early purple orchid, heartsease and the mountain pansy are caught between this and the inner part-border of picks and shovels. The mountain pansy is tolerant of heavy metals and gives a clue to their presence. In Derbyshire this plant has a yellow flower, whereas in the north Pennines it is blue – demonstrating particularity in two very precise ways.

Notes surround a series of façade illustrations taking the observer on 'a 360 degree architectural survey from the Cross, [which] includes many features which appear throughout the village' including, 'sandstone: rusticated quoins. Italian influence in ornate Victorian door. Limestone: random rubble gable end. Staffordshire clay tiles. Gunstock door'. Knitting Frame Workshops are numbered and described. One, dated 1737, is painted with its wide windows, stone slate roof and steps to the first floor with its mention in Pevsner's (1978) survey *Buildings of England – Derbyshire*.

Names lie across street and field: the Clatterway, Yeoman Street, Abel Lane, Sword Lowe, Fairy Ground, The Catsones, Stoney Croft, Upper Wall Sitch, Blackstone Pingle. There are paved walks across the fields from one corner of the village to the other.

From the 'disused dew pond' to the lorry park, with its note 'Site of "Red Triangle Hut" where dances and concerts were held until 1973 when the roof collapsed under the weight of snow', intimations of stories often told mix with fragments of memories: 'Gypsies lived on Arter Hill during WWII – they were

forbidden from travelling'. Another car park has 'site of old calamine mill', a field
has a square enclosure called 'Boiling Pot Reservoir', 'stored silage' rolls up to
Brumlea Farm on Well Head Lane. Wells and fountains are marked in the village
streets. The Pig of Lead – no longer a pub or indeed an ingot of ore – sits in a
quarry which has given up dolerite – locally known as toadstone.

The Via Gellia Mills stand by a dam used by dippers and kingfishers. A skylark
hovers above Bonsall Mines and pigeons, larger than any other feature on the map,
fly homewards above an outline of Europe. Rennes, Nantes, Saintes and Pau are
named – the boundaries of home stretch to faraway railway stations to which the
racing birds are sent for release.

The public WC is marked, the garage on the site of the old gasworks, the
recycling centre, football club, chapels, ridge and furrow, the former CTC hostel
'which had a six seater privy!'. Extinct volcanoes, deep limestone caverns,
fluorspar workings, the lane where Sweet Cicely grows, the village hall (former
school), former shops are marked on land falling from 850 feet to 400 feet above
sea level, close to the Roman road that exploited the mineral wealth of the Peak
and Richard Arkwright's early mills that harnessed water power.

Following the Parish Map, a Village Design Statement (2002) was researched
and written offering careful guidelines for change based in Local Distinctiveness.
Forty villagers met formally over three years and all households were surveyed.
Stone walls, landscape, flora and fauna, historic buildings – are all considered

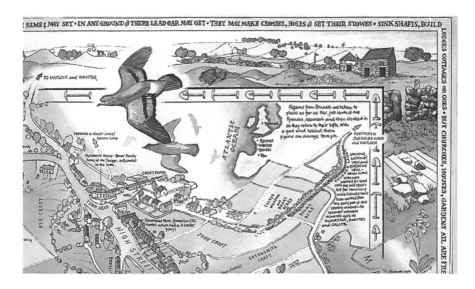

Figure 2.2   Bonsall Parish map

Source: © 1997 Researched and compiled by villagers of Bonsall, designed by Jonathan
Ogilvie, produced as a folded map (with information on the reverse) and a poster

important and pigeon lofts are mentioned as 'distinctive village features, worthy of special attention and conservation by active use for their original purpose'.

West Sussex Parish Maps Project   Kim Leslie, working for the West Sussex County Record Office, saw in Parish Maps the potential for a serious and lasting celebration of the Millennium. From 1998–2001 he ran a county-wide project resulting in 83 maps involving over 1,500 people. Since then almost every other parish has made a map – over 100 exist. His advocacy propelled a flourish of activity and knowledge, sharing and caring that has rolled on and on:

> There is a magical effect on people and communities when they make Parish Maps. In my career I have been responsible for many community projects and I believe this has been the most successful of them all in its benefits to both people and places. Making a Parish Map releases energy. It is about home, community, neighbourhood, issues which are about powerful emotions. A sense of place is a deep and dignifying component of life. I have seen people liberated and made bigger, people have taught themselves new skills and have been given confidence in themselves, communities have been given a new pride and new opportunities. Unlike other maps usually produced for official, business and legal purposes, these are essentially democratic expressions of what people think about the place where they live. They record psychological, emotional and social values. They offer a view of the local world with the potential to involve people in an active way as participants, rather than spectators. (Leslie, K. personal communication)

An exhibition of the maps in Worthing Museum brought thousands of visitors and a magnificent tome has been published including many of the maps with explanatory essays (Leslie 2006). It is sold around the world to activists, academics, ex-pats, and cartophiles.

ABCs

Common Ground has also promoted the making of an ABC as a simple way of revealing the place in a new way. Organizing things by their initial letter has the magical effect of reordering them; it liberates from the expected and routine descriptions. Starting with the basic members of the alphabet, the task begins with the amassing of words of the place amplified by poetry, sketch, photograph, quotation. This levels and reshuffles everything, no longer seeing history as an arrow through time, no longer using old hierarchies; things long taken for granted may be recognized again. Things shared across the country or continents jostle with things vernacular, locally abundant, specific to this place. The assemblage is unique, the fingerprint, the identity of this place in the world.

ABCs like Parish Maps, are ways of getting started in a process of understanding what a place means to you. They welcome anyone to bring their knowledge and their curiosity, to share it and to be enthused into taking their observations into action. An ABC may all be words or it may be decided to create a visual résumé of the place with artists helping where they can. Anything may prove relevant; buildings, details of chimneys, historic texts, people, front doors, legends, trees, local breeds or varieties, foods, dishes, festival days. The initial letters themselves may be gathered locally from gravestones, shopfronts, pub signs, milestones.

Initiated and done by anyone, ABCs can be compiled simply and quickly or lavishly and in depth. Many have provoked interest in Local Distinctiveness and further exploration. Somerset County Council made a poster using photographs. Hampshire County Council have created a folding leaflet with stamp-like illustrations. In Manchester a voluntary organization (The Kindling Trust) have achieved a fascinating city-wide ABC to get people interested in making their own.

The ABC proves a good vehicle for groups and individuals to articulate identity, to orientate, to communicate or even find 'home'. 'A is for Apple' starts the ABC book done by an individual who in her own words was born in Devon, educated in Scotland, England and Canada and after living in Mexico, Austria and India moved back to Britain. 'There were several things that I didn't understand when I first arrived in Southwell: what a huge church was doing here; what a prebend was, or "wong", "burgage", "dumbles" … So I looked things up, and pestered people and then wrote this'. This Nottinghamshire town is the home of Bramley's Seedling, the original mother tree still standing after 200 years in the garden of a terraced house (Tapply 2006).

Bath has a persona known to most people through its Roman Baths, Georgian architecture and the writings of Jane Austen. The Museum of Bath at Work opened a new gallery in 2007 specifically designed to provoke and offer complementary perspectives of the city. The Hudson gallery now houses 'Bath in Particular' – a permanent but changing exhibition prompted and filled by local participation. Eight local historical societies guided by the idea of Local Distinctiveness and ABCs provided their own view of their places within the city. Inviting museum users and visitors to suggest and offer materials for inclusion makes the process interactive, unpredictable and exciting, enthusing people to look at these and their own surroundings with new eyes. Walks and talks are just some of the spin offs.

Our friends in Italy have also used the ABC as a way of rousing themselves and visitors to appreciate the rich mix of place (Figure 2.3). For Cortemilia the valley mists, ravioli del Plin, a nineteenth-century brigande, family names typical of the area, stone terraces, working masons, and bridges are just some of the bold illustrations.

Working for English Heritage in Yorkshire, Keith Emerick has found the challenge of making ABCs brings out of the community knowledge and action that the professionals could never have achieved on their own.

In our day-to-day work of managing those monuments considered to be 'nationally important' English Heritage Yorkshire Region is using some of the

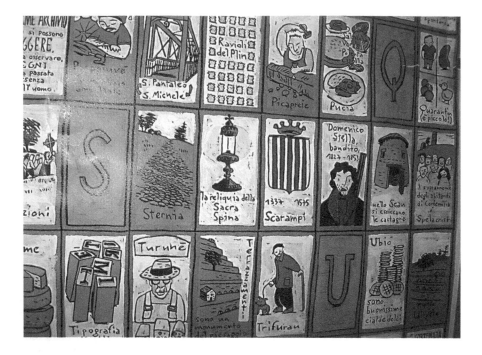

Figure 2.3   Detail from Per celebrare la specificita locale, Ecomuseo dei Terrazzamenti e della vite, Cortemilia, Regione Piemonte, Italia 2000

Source: © Words: Donatella Murtas, design: Marto Martis

techniques of local and community engagement familiar to Common Ground. We believe that for the sustainable management of a 'place' to be achieved it is essential that every effort is made to understand the multiplicity of values attached to that 'place' – not just those values imported by the heritage experts. Our aim is to get a better understanding of ideas of 'value' and 'meaning'.

At Cawood, near Selby, North Yorkshire, English Heritage has promoted the use of the formal 'Conservation Plan' process, but we have also successfully suggested to the residents that they prepare 'ABCs' of local significance. Cawood is a large village with two substantial Scheduled Monuments, both of which are of medieval origin and now large open areas of grass, earthworks, old orchards and fishponds. The village purchased one of these – Cawood Garth – in order to use it as community space. However previous approaches to the management of Scheduled Monuments were loathe to accept change or the use of such places, and there was thus a curious mix of something being considered 'nationally important', being managed for the 'public good', but the public were always unable to use, or were excluded from the site. There has now been a change of approach with participation and inclusion the dominant themes – although I would be the first to admit that there has not been a consistent application of this approach. At Cawood we suggested that the residents should write their own Conservation Plan for the Garth – rather than have it prepared by consultants parachuted in – and after initial reservations the residents have thrown themselves into the preparation of this document, with the committee responsible for it now the most active and sought after position in the village. The Plan has now become about the village rather than just the Garth – which, as far as we are concerned, means that the Garth has a real, living context. As we hoped, some fabulous stories have emerged: a live beluga whale turned up in the river one year; another year two porpoises appeared in the river and stayed for the Christmas period; Cawood was the setting for one of the biggest meals ever served in England with several hundred cooks brought in by the Percy's for one celebratory meal in the Middle Ages.

The Conservation Plan approach to managing places hinges on defining 'cultural significance', but because this concept can be difficult to explain we used the idea of the 'ABC' as a way of getting people to talk about those things in the village that were important to them. And, again, once the idea was presented to the village they took it up and made it their own. The adults produced one 'ABC' and the primary school children produced another. Both the Conservation Plan and the 'ABC' work well together because the 'ABC' provides the local significance, but it also illustrates how significance is always changing which will make it easier to adapt the Conservation Plan when that is ready for renewal. The outcome of all this work is that the community has a document that they own and which can guide their use of their space.

We realise that there is a considerable tide of past experience to turn back before people accept us as 'enablers' rather than policemen, but the promotion, encouragement and above all legitimation of community initiative is in our view the only thing that can make 'monuments' sustainable and we are grateful for the advice and support of Common Ground. The 'ABC' fold-out of local distinctiveness has been invaluable, and in our short experience of promoting it, it has captured the imagination more than any other tool.

(Keith Emerick, Inspector of Ancient Monuments, North Yorkshire and Neil Redfern, Inspector of Ancient Monuments, West Yorkshire. February 2005, personal communication)

Local and Universal

People have been in touch with us from across the world interested in Local Distinctiveness, Parish Maps and ABCs. We have spent time in North America and Australia where these ideas have spread. In 2002 Common Ground were invited to work with the Instituto di Recherché Economico-Sociale del Piemonte based in Turin and in concert with eco- and community-museums across the country. With Kim Leslie's help, many there were beguiled by the potential and since that time the spread of interest owes much to his repeated visits. And in West Africa and Eastern Europe Mike Flood with Powerful Information has used Parish Maps with many communities. Local Distinctiveness has applicability and relevance anywhere when guided by local culture.

Accumulation and Assemblage

Respecting the accumulations of story upon history upon natural history that give a place its uniqueness, is good grounding for creating a new relationship with nature.

Story

The details of a place spark the telling of stories, act as lightning conductors for both rousing curiosity and passing on knowledge. Stories retold keep a place and its community in touch with the depth of its past.

A pool, now measured at over 100 feet deep, carries a name and a tale from Saxon times as told on the Lyminster map (West Sussex). Amongst the many details, a tower incorporated into Priory Farm, Crossbush Lodge with its Sussex flint work, the Black Ditch, the Driftway and the Knucker Hole. In this 'bottomless pool', the story goes, lived a predatory dragon. The dragon was finally brought down by a poisoned pudding gaining the young man whose plot succeeded the hand of the daughter of the local lord. But in the celebrations at the Six Bells, poor Jim Pulk wiped poison residue across his mouth and died on the very same day.

The Copthorne map (West Sussex) itself tells a many stranded story. Kim Leslie describes the map as resembling:

> One massive oak tree. Its roots are deeply embedded into the landscape under a huge green canopy bearing its harvest of acorns. This is the tree that's a map that's an expression of community life – a brilliant conception by the map-makers of Copthorne ... the most typical of Sussex trees, so prolific that Kipling called it the 'Sussex weed'. To the people here the oak symbolizes the strength of community spirit within the village. And it is the tree that creates the map. Roads, footpaths and boundaries form its branches, giving it shape and form. The tangled and winding roots feature the family names of everyone living in the village – there are close on fourteen hundred names written here ... (Leslie 2000)

The place is its people, the tree as metaphor, the map embedded in its branches, the people safe in its roots. And this old, originally isolated woodland community, with its tales of smuggling, poaching and fist-fighting is standing up against boundary issues between Sussex and Surrey and the ever expanding Gatwick Airport.

Again in West Sussex, Tangmere's pub is called the Bader Arms, so it should come as no surprise that the map is dominated by a Spitfire, with many other man-made 'kites' in attendance. Tangmere Airfield Nurseries is a remnant in name of the RAFs presence. Street names still remember other planes and wartime air crew. The land requisitioned for wartime was finally sold off in 1979. Since then the population has tripled, light industry and housing taking over the airfield. The map tells the story of change and change and change again all in a single century.

The Manchester A to Z guide by the Kindling Trust (2008) is full of interest (Figure 2.4). The smooth, poised Manchester terrier, 'a cat-like dog bred to hunt rabbits, rats and mice' stands next to an easily recognized hirsute Karl Marx with a note mentioning that Frederick Engels brought him to the city in July 1845. The serendipity of their juxtaposition sets the mind wondering what they have in common. The poverty of working class housing in the city shocked Engels and influenced Marx. In those conditions a dog bred to catch a meal and to keep you free from rats makes sense.

Figure 2.4   Detail from Manchester in Particular: A to Z guide to Manchester's unique political, cultural, ecological and economic landscape

Source: © Kindling Trust 2008

History

Unwritten history taunts us with every change of light and season. The things we pick up from the turn of a street, or the shadow in a field reinforce the importance of placing oneself in continuing history, understanding the dynamism of place. Names from languages long forgotten populate our fields and cities, rivers and hillsides, buildings and trees; they murmur other lands. From where does the name Chideock hail? The Dorset map gives 10 spellings of the name to muse over. And what may be hidden in its field names: Frying Pan, Brick Field, Water Sheares, Roundy Tweazles, Great Olivers, Long Lands, Worse Again, Flery Pins, and Yennidge.

So often the visible and invisible persistence of economic activity is revisited on the maps. Combe Martin's map is bisected by one of the longest villages not only in Devon, but in the country. It follows the River Umber and makes mention of iron, manganese and silver mines 'worked intermittently between 1293 and 1880 ... the main ore was silver lead galena which was said to have sparkled like the inside of a jewellers shop by the flickering light of the miner's candles'. The people who made the Parish Map of Tow Law in County Durham with its mix of mining and agriculture would empathize with the cultural as well as landscape changes, here more recently felt.

Edge, boundary, border – with their implications of parochialism, exclusivity, otherness – can bring out the worst in humanity. Identity within a small ancient 'kingdom' can be keenly felt, but positive parochialism seeks to connect not to divide. The feeling of home, in Welsh the words *cynefin* and *bro*, in German *heimat* carry strong associations. There is no equivalent word in English. Not always, but often the parish boundary is taken as the edge of things for a Parish Map. The ancient ecclesiastical boundary and the more recent civil parish sometimes break step but the marching of geography with history often brings richness in wild life too (boundary banks offer a good example) and the tracing of old edges and hedges can bring disciplines and community together.

In Averton Gifford (Devon) research parties wandered the parish often retiring later to the pub. Conversations led to photographic workshops resulting in an exhibition of rare archive pictures which prompted stories and memories. Since the production of the original map, the group has made walk leaflets, the revival of the beating of the parish bounds have been celebrated with Rammalation biscuits and Ganging Beer made by local businesses, and they have promoted the restoration of historic buildings in the village for public use.

A group of artist publishers (Two Rivers Press, Reading) decided in 1995/6 to explore the Ancient Boundary of Reading (Figure 2.5). The wordsmiths placed stories running in all directions on a beautiful printed map of their 'parish' (three actual parishes). In just two colours and almost completely in words, they trace the stories and names, features and relics that define the old settlement.

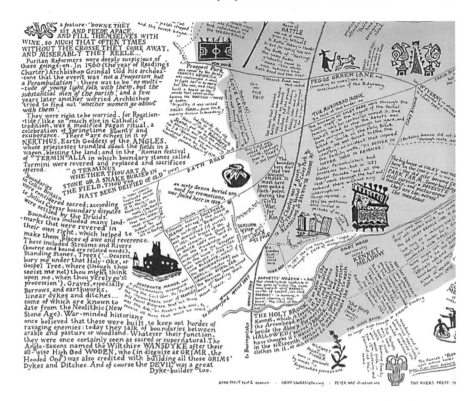

Figure 2.5   Detail from Boundary of Reading
Source: © Two Rivers Press 1996

Three, maybe four boundary burials so far discovered show that the Boundary's significance is at least as old as they are and probably a good deal older still: the limits of an Iron Age 'estate' which contained enough land to support a small community? The line can still be traced – a marker here, a wall or ditch there, or a gap between the houses.

The map is packed with information, stories and observations that animate the jostle of history with glimpsed asides from centuries of perambulations. From 'Milkwort or rogation flower worn in garlands by maids in rogation processions' to the man who traversed 'the spring line of ponds now filled in – 'the black mud had to be literally scraped off him' (1861)'.

Gartmore near Stirling created their map depicted as though through a fish-eye lens. At its centre lies the formal village main street with details and names of fields and village houses, but it is contained in a circle of horizons – Ben Lomond 3,192 feet (six miles distant), Meikle Bin 1,870 feet (13 miles), Dumyat 1,373 feet (19 miles), Campsie Fells (9 miles). The things cherished here catch and frame this floating world – the local and the universal understood in a graphic and powerful way.

Natural History

There are many starting points and however driven people may be by one aspect of their place such as buildings or butterflies, somehow there are always connections and especially with the geology and topography, the flora and fauna, the sky.

The people of Grappenhall (Cheshire):

> quickly realized that our parish is conspicuous by its lack of scenery, that is, everything seems to be about 20 feet above sea level. What we do have, however, is water, lots of it, and vast expanses of sky. The flat terrain does make us all too aware of the weather and so we are very much aware of the passing seasons. We used these aspects as the background, looking for contrasts and comparisons: ploughed fields ... crops and harvest ... Bridgewater canal ... the Walking Day procession ... the shire horses ... annual steam fair ... the church ... The cobblestones ... Manchester Ship Canal ... the tiny stream the Morris Brook ... (Cheshire Landscape Trust, personal communication)

Elham's Parish Map in Kent is an 8 foot wide triptych. Added to the portraits of all the houses (and Les Ames, cricketer, born here 1906–90) there are mammals including moles, fungi including fly agaric, trees (sweet chestnut and hazel), beetles, flies, birds and flowers, butterflies and domestic animals, walks (more than 50 footpaths have been tended), ancient hedgerows and trees such as 'Grandmother's tree'. The Chalk Pit became a focus of interest, to the extent that dumping was stopped and a reserve for chalk-rich flora defined.

A nine mile stretch of the river Wear is rendered in textile: sand, starfish, shells, sea walls and waves mark the mouth created by one of the 21 schools in Sunderland in a joint map. After a triumphant tour of the schools, the map moved to a permanent home in the city museum. Students who worked on it now return, as young parents, to show their own children.

Liberated from traffic by a bypass, Averton Gifford (Devon) reclaimed their surroundings by walking and making a map. An artist contributed with a class of 8–10 year olds. Grown-ups were reminded they too played in the stream, excited as the children by silvery shimmering pebbles which when touched disturbed tiny clinging elvers. 'The face tree' (traceable in its bark) with low limbs good for climbing and,

> hollow hedgerow trees where you could lose your arm right up to the shoulder, were explored along with drain covers and paving slabs which made patterns of flowers and diamonds, and a collection of derelict barns which were haunted. Certainly ghostly giggles could be heard as soon as we got near! Quick sighted

and sharply observant, the children could teach many a long-standing resident to
see the village with new eyes. (Sally Tallant, personal communication)

This with much more information was entrusted to Mike Glanville who, over long
winter evenings began to shape a map to be printed. His watercolour is embellished
with wild creatures and flying visitors – the north point is a welcome swallow
arriving. There are identifiable species of fish in the tidal river, a full portrait of the
vital bridge and the warning poles along the twice-daily inundated lane.

Wild life enthusiasts started the Selsey Map (Sussex) as a way of drawing
others into an appreciation of nature close to them. It includes the lifeboat station,
tornadoes, mention of the sounds of the bells of Wilfrid's cathedral reputed to
be heard ringing below the sea on stormy nights – all reminders of the treachery
of the sea and the weather hereabouts. Cottages made from railway carriages,
caravans and crabs, field names, fish, birds of the land and sea, flora and fauna
animate this map.

Combe Martin (Devon) is recognizable from the unique combination of heron,
sessile oak and red stag looking out from the moor over the colourful hobby-
horse central to the re-enactment of the hunting of the Earl of Rone – all this
amongst illustrations of gorse, heathers, butterflies, starfish, cliffs, blue lobster,
medieval strip fields, scones, cream, strawberries and a warming note: 'the annual
strawberry feast was the highlight of the year'.

Closing Thoughts

Nature constantly explores, backtracks, reacts, negotiates, changes. Places too must
be dynamic, but change needs to be decided in the full knowledge of who gains
and who loses as well as what is lost and gained. We have come to understand that
simply arguing for the wild or the old is not as potent as demonstrating the richness
of the interactions where we and nature have invented together a relationship which
is long-lasting – in other words where we have learnt and practiced sustainability
over a long time.

This demands community, something that has been mislaid in so many places.
But ABCs and Parish Maps have a good track record here:

> Not surprisingly the Map has not changed everyone's idea of whether or not
> Barrow is a great place to live. At least one family was disappointed by the
> lack of photographs of beautiful Barrow. But is it a coincidence that for the first
> time we held an official switching on ceremony of the Christmas lights with
> community carol singing? ... That we are negotiating to buy an old forge and so
> bring it into the public domain? ... The effects of making the map can be seen in
> the village design statement, the newly opened fossil trail, the enthusiastic panto
> group – the Map gave the village a sense of the possible, and marked an urge to

look forward with hope, and not simply back with regret. (Helen Sadler 1997 and 1998, personal communication)

The Cheshire Landscape Trust and the county Women's Institute provoked over 30 maps in 1992/3. In Grappenhall the local WI decided to make a photo collage: 'we found it a mammoth task, what to put in, what to leave out ...' These, of course, are the defining questions for the Ordnance Survey, for policy-makers, for scientists. And however formal or informal the gathering of information – surveys, questionnaires, interviews, desk studies, satellite observances, field work – after all the studies have been done, the other question that still remains is, 'what is important and to whom?'.

Local people may well choose to describe their place quite differently, in ways that confound or complement the findings of experts. After all they are reclaiming their previously unacknowledged interest in the land, they are asserting a communal hold on their surroundings that they have previously only felt individually and hopelessly. Whoever makes the map has the power.

The challenges that we face are greater than ever before. Our need for community resilience and living well with Nature will not be solved by one exclusive blueprint. Local Distinctiveness offers a way of exploring and expressing *place* as the ground truth of long practiced human endeavor with Nature. Parish Maps and ABCs can help in the struggle to help people communicate what their place means to them and to generate and develop community conscientiousness in the face of a relationship with Nature that is breaking down. As we say in *England in Particular*:

> Nature will endure whatever our actions bring. It is we who are in danger. We deprive ourselves of a rich life. We need to live better with the world, and it is our ordinary actions that will be our salvation or our downfall. To ground ourselves, understand our place, find meaning and take steps to cherish and enrich our own patch of land demands that we change our ways, share our knowledge, get involved. We have to know what is of real value to us, where we are, and find new ways of belonging. (Clifford and King 2006: xiii)

References

Aberley, D. (ed.) 1993. *Boundaries of Home: Mapping for Local Empowerment.* The New Catalyst Bioregional Series. Gabriola Island BC, Canada and Philadelphia, USA: New Society Publishers.

Baker, D. 1983. *Living with the Past – The Historic Environment.* GB: David Baker.

Clifford, S. and King, A. (eds) 1993. *Local Distinctiveness: Place Particularity and Identity.* London: Common Ground.

Clifford, S. and King, A. (eds) 1996. *From Place to PLACE: Maps and Parish Maps*. London: Common Ground.

Clifford, S. and King, A. 2006. *England in Particular, a Celebration of the Commonplace, the Local, the Vernacular and the Distinctive*. London: Hodder & Stoughton.

Clifford, S., Maggi, M. and Murtas, D. 2006. *Genius Loci: perché, quando e come realizzare una mappa di comunità.* StrumentIRES 10. Torino: Instituto di Recherché Economico-Sociale del Piemonte.

Common Ground 1994. *Celebrating Local Distinctiveness*. London: Common Ground/Rural Action.

King, A. 1980. *Paradise Lost*. London: Friends of the Earth.

King, A. and Clifford, S. 1985. *Holding Your Ground: An Action Guide to Local Conservation*. 1st Edition. GB: Temple Smith.

Leslie, K. 2006. *A Sense of Place: West Sussex Parish Maps*. Chichester: West Sussex County Council.

Shoard, M. 1980. *The Theft of the Countryside*. GB: Temple Smith.

Tapply, S. 2006. *S is for Southwell*. GB: The Handmaid Press.

www.commonground.org.uk.

www.england-in-particular.info.

Further Reading

Crouch, D. and Matless, D. 1996. Refiguring Geography: The Parish Maps of Common Ground. *Transactions of the Institute of British Geographers New Series* 21(1), 236–55.

Hayden, D. 1995. *The Power of Place: Urban Landscapes as Public History*. Cambridge, Mass.: The MIT Press.

Lippard, L.R. 1997. *The Lure of the Local: Senses of Place in a Multicentred Society*. New York: The New Press.

Wood, D. 1992. *The Power of Maps*. New York: Guilford Press.

Wood, D. Forthcoming. *Rethinking the Power of Maps*. New York: Guilford Press [includes 'The Outside Critique: The Parish Maps Project'].

Parish Maps Mentioned in Order of Text Appearance

Thirsk Civic Society, North Yorkshire, 1989. [People of 5 parishes with artist Margaret Williams.]

Muchelney, Somerset, 1987. [Gordon Young, artist.]

Bonsall, Derbyshire, 1997. [Researched by villagers, designed by Jonathan Ogilvie.]

Grappenhall, Cheshire, 1993. [Local Women's Institute.]

Elham, Kent, 1994. [The Elham Circle.]

The River Wear Map, Sunderland, 1989. [Schools of the city, Tyne and Wear Museums.]

Averton Gifford, Devon, 1992. [Researched by parishioners, artist Mike Glanville.]

Selsey, West Sussex, 2001. [Produced by Joe and Eileen Savill, Jean Bankes, Sarah and Peter White.]

Combe Martin, Devon, 1994.

Tow Law, County Durham, 1990.

Chideock, Dorset, 1990. [Leader Kate Geraghty, painter Gillian M. Moores, Dorset.]

Ancient Boundary of Reading, 1996. [Two Rivers Press Berkshire. Peter Hay illustrations, Adam Stout research, Geoff Sawers lettering.]

Lyminster, West Sussex, 2000. [Team leader and designer Alan Burnett.]

Copthorne, West Sussex, 2000.

Tangmere, West Sussex, 2000.

Barrow upon Soar, Leicestershire, 1996.

ABCs Mentioned Likewise

Somerset County Council, Sustainable Somerset Group, 1995.

Hampshire County Council, Sustainability Team, 2007/8.

Manchester in Particular: A to Z guide to Manchester's unique political, cultural, ecological and economic landscape. Kindling Trust 2008.

Per celebrare la specificita locale, Ecomuseo dei Terrazzamenti e della vite, Cortemilia, Regione Piemonte, Italia, 2000.

Chapter 3

Marketing Sense of Place in the Forest of Bowland

Cathy Hopley and Paul Mahony

Introduction

The Forest of Bowland Area of Outstanding Natural Beauty (AONB) is a protected landscape in upland Lancashire and North Yorkshire (northern England), which is internationally important for its heather moorland, blanket bog and rare birds. Recent work to develop sustainable tourism in the area identified the need to create a 'sense of place' or 'brand identity' for Bowland in order to raise awareness of the area and its special qualities, and to encourage tourists to develop a loyalty to the area, and so become regular visitors. This was achieved through the creation of a sense of place toolkit and provision of training courses for tourism businesses and other partners, and the development of a brand and communications strategy for the AONB Unit itself. The toolkit has become widely known and respected within the sustainable tourism sector and protected areas network in the UK, with similar work now being developed in the Cairngorms, South Pennines and the Clwydian Hills.

For the purpose of this work, sense of place was interpreted as 'an area's unique feel and appearance; made up of the different landscapes, wildlife, heritage, people, sights, sounds, tastes, memories and many other elements that form our relationship with a place.'

We decided to use this relationship with place as a marketing tool to achieve two clear aims:

- to raise awareness of the Forest of Bowland AONB as an entity, and to improve understanding of the area's special qualities by providing clear, consistent messages about the area (this would generate a 'brand identity' related to the place); and
- to develop an affinity and loyalty to the area by both residents, businesses and visitors (the beginnings of a 'destination brand' for the Forest of Bowland AONB).

This marketing was to be carried out both by the AONB unit itself via its website and limited promotional work; and, perhaps more importantly and effectively, via the partners and businesses who promoted their own products (accommodation, food, activities and attractions) within the developing 'brand' of Bowland.

Understanding Sense of Place

Our research into a sense of place began with interpretations from geographical studies – the term sense of place being coined as a reaction to the feeling of 'placelessness' identified by Relph (1976) and 'clone town' developments in the US and UK (Simms et al. 2005). We were strongly influenced by the work of Common Ground (see Chapter 2) and their development of 'local distinctiveness' where each place has its own character developed by people and the place over time. Their work on parish maps was especially important to us, as was the developing ideas of an 'A to Z' of local distinctiveness. Working with local people in a distinct community enables you to make these connections, and that is what our work was seeking to build on in order to support sustainable tourism in the area.

Recent work led by The Research Box, Land Use Consultants and Rick Minter, on behalf of Natural England (2009) has sought to explore the 'cultural services' which landscapes can offer to people's quality of life (as part of a wider investigation into the 'ecosystem service' offered by the natural environment). This identified eight cultural services, which included inspiration, escapism, a sense of history and a sense of place. It suggests that people have a portfolio of places which they use for a variety of experiences – using local accessible places for regular stress relief, and more special places for more intense experiences. They found that any feature, landscape, or even iconic wildlife, could be considered to deliver a sense of place if it was local and distinctive to the area. People were 'proud' of such places, because of their history or because they defined a 'mood' of the local landscape

We were also influenced by a previous 'toolkit' approach pioneered by the Wales Tourist Board in 2005 which sought to identify key themes of place and to encourage businesses to research and use these in their local situation and in their marketing. Their work focussed on Wales and its people; using the Welsh language (an added dimension of 'place'); working with buildings inside and out; food and drink; using creativity and the arts; and the great outdoors. As they pointed out, a sense of place makes sense:

- it is practical – it is a resource that is all around you;
- it is financially rewarding – it adds value to your business, often at little or no extra cost;
- it is inspirational – it can unlock ideas and innovation;
- it is friendly – it offers an opportunity to work with others and with the community; and
- it is common sense!

Background to the Forest of Bowland AONB

The Forest of Bowland AONB is an area of over 300 square miles which is managed by a partnership of landowners, farmers, voluntary organizations, wildlife groups, recreation groups, local councils and government agencies who work together to protect, conserve and enhance the natural and cultural heritage of this special area (see Map 3.1).

The Forest of Bowland AONB was created in 1964 with a primary purpose to conserve and enhance the natural and cultural heritage of the area, whilst having regard to the social and economic needs of the landowners, farmers and communities. A secondary role is to encourage enjoyment of the area, but only where this is consistent with its primary purpose. In this way AONBs (there are 49 in England, Wales and Northern Ireland) differ from National Parks which have a clear remit to 'promote opportunities for the understanding and enjoyment' of the area and have therefore had a longer and stronger tradition of tourism, a well known place identity and a higher profile as a destination to visitors.

The 'Forest of Bowland' is not particularly well known in the North West of England. As an area it is made up of several better known places and destinations – such as Pendle Hill, the Trough of Bowland, Ribble Valley, Lunesdale and various other fells, moors, valleys and villages. Only one small part of the AONB has traditionally used the Forest of Bowland or Bowland Forest nomenclature and this is an area which was originally the Royal Forest of Bowland, owned and hunted by the King in medieval times. Getting people to relate to the Forest of Bowland place name was therefore the first challenge.

A second challenge was to raise the public's understanding of the AONB designation, area and special qualities which was also recognized, from various surveys and consultations, to be fairly low. In the tourism sector in particular many businesses promoted themselves as being 'close to the Yorkshire Dales and the Lake District' and would encourage staying visitors to go out of the area for day trips as they knew little about what was on offer in Bowland itself. We were aiming to develop sustainable tourism in the Forest of Bowland AONB and it was therefore vital that we encouraged visitors to stay in the area, explore it, spend their money there, and want to return to discover more.

**Map 3.1   The Forest of Bowland Area of Outstanding Natural Beauty
located in the North West of England**

Source: Map data © OpenStreetMap contributors, CC-BY-SA

Sustainable Tourism

Sustainable Tourism takes its lead from sustainable development, a term which has become increasingly prominent since the Rio Earth Summit of 1992. Sustainable development can be defined as 'development that meets the needs of the present, without compromising the ability of future generations to meet their own needs' (United Nations World Commission on Environment and Development 1987). With this thinking in mind sustainable tourism aims to make a low impact on the environment and local culture, while encouraging better income and employment, and conservation of the very landscape upon which tourism is based.

Within the Forest of Bowland AONB there are many sensitive and valuable habitats and species. There is a need therefore to ensure a balance that allows these areas to be carefully managed and protected from the pressures of visitors and development, whilst also enabling and encouraging these visitors to experience the beauty of the area, and to allow communities and businesses to be viable and successful.

Tourism as a sector of the rural economy in the Forest of Bowland AONB had traditionally been small; the area had long been popular with day visitors from the surrounding urban areas of East and North Lancashire (one million people live within a 30 minute journey of the AONB), but staying visitors were not well catered for, and income generated by visitors was low. This situation began to change in the late 1980s and 1990s and, after the unwelcome impact of Foot and Mouth Disease in 2001 when many farmers lost their entire stock, more businesses looked to diversify into tourism to generate a living and stay in farming. The AONB Unit was keen to support this shift believing it could create a more sustainable future for the area. Therefore the AONB partnership worked hard to develop leadership in the sector by consulting widely and publishing a strategy for sustainable tourism in the Forest of Bowland AONB in 2005. This was quickly followed by an application for, and the awarding of, the European Charter for Sustainable Tourism in Protected Areas – the first in England.

Working from this base (as the strategy had recognized the two key challenges of a low level of recognition of the Forest of Bowland place name, and the lack of understanding of the designation, area and special qualities amongst businesses) the AONB Unit looked to develop the Sense of Place Toolkit.

Developing the Forest of Bowland Brand

The term 'branding' is typically associated with consumer advertising; we tend to understand it in terms of the various logos, slogans, jingles and other means through which products and services are sold to us. Such brands exist to differentiate products from their competitors and capture the attention of potential customers, with the ultimate aim of persuading people to 'invest' in the products – both financially and emotionally.

Branding and marketing products in this way is certainly nothing new and perhaps as old as commerce itself. Yet the application of these principles in branding and marketing rural areas remains largely underdeveloped. This is partly due to a lack of investment, but also on account of traditional marketing efforts having relied on a more homogenous approach to promoting rural areas – whereby the majority of rural destinations now share the same or similar brand identities and tourism offers, with each competing to shout the loudest in an increasingly crowded marketplace.

The consequence is that the rural landscape, although rich in diversity and character, is often generalized through marketing activity and portrayed as a somewhat uniform and even 'placeless' environment. Browse through the leaflet rack in any rural Tourist Information Centre and you will encounter the same imagery and messages time and again: a generic stereotype of the countryside as being little more than stunning scenery, picturesque villages, rare wildlife and so on; a nostalgic portrait of rural life as some form of living museum. While this image is perhaps a positive one and undoubtedly ingrained in our perception of the countryside, it remains a shallow interpretation of what today's rural landscape has to offer.

The challenge is therefore for destination managers (and all stakeholders) to step back from the rural stereotype and focus instead on understanding and celebrating what is special, unique or simply different about their local landscape. It is here where the principles of branding and target marketing can truly add benefit, providing a vital means by which destinations can differentiate themselves from one another and express their sense of place – not just for the purpose of attracting visitors, but also as a way of strengthening communities and bolstering people's pride in their local area.

In the Forest of Bowland AONB, an understanding of sense of place has been used in this way to successfully re-brand and reposition the area's marketing effort. This has involved revision and development of both the area's brand identity and its overall destination brand – two similar yet distinct concepts that we will now briefly explore for the purpose of clarity.

An area's brand identity and its destination brand are closely interlinked, although there is an important distinction to be made between the two. A brand identity is all about the 'look and feel' of how a place is promoted and portrayed. Brand identities are not so much a product of the place in itself, but are instead a product of the custodians responsible for managing it, most often for the purpose of promoting a place to visitors (for example, the Bowland brand does not belong to the landscape itself, but is instead the property of the AONB partnership). A brand identity serves to ensure consistency in how a place is presented and portrayed.

A destination brand is all about the reputation of a place as perceived and experienced by local communities and visitors; it is what a place is known and recognized for, the *reason* why someone may choose to visit. A destination brand is not a single, tangible product but an amalgamation of the various features that make a place special or unique, much like a sense of place. Unlike a brand identity, which can be created and put into effect relatively quickly, destination brands are notoriously difficult to influence and even then require many years to develop into fruition. This is because the essence of a destination brand lies in hearts and minds of local communities, businesses, visitors and other stakeholders, which cannot be so easily shaped and controlled as a logo or publicity campaign.

In the case of the Forest of Bowland AONB, the area's brand identity and overall destination brand were reviewed following an extensive 'sense of place consultation'. Opportunity was seized to use the understanding gained through consultation to revise the AONB's identity in line with people's own experiences and perceptions of the area.

The outcome of this process was the development of a series of interpretive themes, or 'sub-brands', each one representing a special quality of the area to which both local people and visitors attribute great value and attachment. Each theme was given a title or slogan, based upon the special quality to which it relates, as listed below:

- *A Place to Enjoy and Keep Special:* the over-arching message of the AONB. It combines the importance of the area for people's livelihoods and enjoyment, with its value as a unique and protected landscape.
- *Delicious Local Food and Drink:* celebrating the wealth of high quality foodstuffs produced in Bowland and the role this plays in contributing to local culture, traditions, livelihoods and people's sense of place.
- *Wild Open Spaces:* showcasing the character, quality and beauty of the Bowland landscape. And promoting this not only as backdrop for recreation, but also as a way of adding value to local products and services.
- *A Landscape Rich in Heritage:* exploring the wealth of local culture and heritage in the Bowland landscape. Shifting the spotlight away from established landmarks to illuminate the hidden histories that lie behind local stories, places and traditions.
- *A Living Landscape:* understanding the role played by people in shaping the character and sense of place of the landscape, both past and present. From the people who farm and manage the land, to the vibrant community and interest groups and events that play an important role in welcoming visitors.
- *A Special Place for Wildlife:* highlighting the distinctive wildlife of Bowland, its rare and protected species as well as the more common yet characteristic plants and animals that can contribute to a unique and memorable experience for visitors.

It would be fair to say that all of these themes are somewhat generic and could be used to describe many rural destinations. However, when applied to the Bowland brand, they offer a framework through which the AONB's marketing effort can be delivered more effectively. This is a simple step perhaps, but an important one in helping to shift the focus of AONB publicity away from promoting the area as a whole and towards a more targeted approach, focusing on those aspects of the destination brand (or sense of place) that customers most identify with, or are most interested in.

The next step in the re-branding process was to focus on evolving and expanding the existing brand identity to accommodate the newly developed themes, or sub-brands. There was a need for each theme to have an identity of its own, enabling it to become more easily recognized by its target audience. Yet it was also vital for each theme to remain very clearly a part of the overall Bowland brand identity, which at the time was less established and vulnerable to any potential dilution or overhaul.

With this in mind, the necessary distinction between each theme was achieved simply through use of colour and imagery. Themes were colour-coded individually or in pairs, with each colour representing a broad topic of interest for potential visitors, as follows:

- Green for 'Wild Open Spaces' and 'A Special Place for Wildlife', targeting people who like to get outdoors and experience nature or simply take in the view.
- Burgundy for 'A Landscape Rich in Heritage' and 'A Living Landscape', targeting people who like to learn about the histories and cultures of the places they visit.
- Blue for 'Delicious Food and Drink' – this colour has also been adopted by the AONB's Sustainable Tourism Business Network and is now used to brand all publicity materials targeting people with an interest in local products and services generally (i.e. not just foodstuffs) (see Figure 3.2).
- Orange for 'A Place to Enjoy and Keep Special', the AONB's original brand colour, which is now used exclusively to promote recreational activities, as well as headline publications such as the AONB annual report (see Figure 3.1).

When applied to Forest of Bowland AONB communications, this simple system of colour-coding enables visitors to more easily identify and access information that is tailored to their interests, be it a colour-coded leaflet, web page, event banner or other material.

The Forest of Bowland Visitor and Enterprise Surveys 2005–10 show that this process of involving local stakeholders in the frontline of tourism marketing has proved successful, as demonstrated by an increase in the number of visitors surveyed who stated that they saw the AONB as a 'destination', from 21% in 2007, to 33% in 2008 following the re-branding exercise. A total of 88% of Bowland businesses also now market themselves as being 'within an Area of Outstanding Natural Beauty'.

Figure 3.1 **The Taste of Bowland booklet, in blue, lists businesses which produce and sell food in the area**

Source: Forest of Bowland AONB

**Figure 3.2   The Bowland by Bike leaflet, in orange, suggests a number of
routes for exploring the area by bike**
Source: Forest of Bowland AONB

The Wider Context of Re-branding Bowland

The Forest of Bowland AONB is not alone in re-positioning its brand to focus attention on valuing 'the local'. Many other rural destinations are now beginning to base their tourism offer on an understanding of sense of place and local distinctiveness. Furthermore, this trend is also apparent in some areas of the wider economy, as consumer concerns over climate change issues (particularly so called 'food miles') and the traceability of products create new opportunities for adding value to goods and services based on their heritage and place of origin.

A notable example is the case of premium foodstuffs, which have seen a recent boom in popularity in response to increasing public awareness of food quality and associated impacts on health. Such awareness has been partly driven by national initiatives to encourage healthy eating, as well as negative publicity surrounding poor practice in food production (from the BSE outbreak in 2000/01, to the more recent vilification of battery-farmed and processed foods by mainstream media). As a consequence, the link between food and the landscape is undergoing a resurgence, with consumers becoming more proactive in seeking out products they perceive as being natural, wholesome and authentic. Knowing *where* our food comes from and even who produced it is now an important factor in our understanding of food quality, and this is mirrored by the way in which premium foodstuffs are branded and marketed.

For instance, browse the aisles of any major supermarket and you will find premium products with their place of origin emblazoned across the front of the packaging, often as part of the brand name itself. In some cases you will also find a photograph of the producer, most likely stood on his or her estate so that you can literally see the place from which the product has originated. In some cases, marketers have even created fictitious places in order to add a sense of authenticity to their products, one example being Marks & Spencer Lochmuir Salmon and Oakham Chicken, both of which are brand names designed to sound like real places, when in fact the products sold through each brand are sourced from across Scotland and the East of England respectively. That is not to say that Lochmuir Salmon and Oakham Chicken are in any way inferior products because of this, but rather that consumers' association with place is clearly strong enough to influence the development of such brands.

So how does this relate to the Forest of Bowland AONB? The key point is that the re-branding of Bowland ties in with an emerging trend that can be found throughout not only the tourism market, but also significant areas of the wider economy: that of adding value to products, services and experiences by portraying them as being authentic, traditional and characteristic of a place. Clearly, these are all qualities that the Bowland landscape has in abundance and which can be used to tap into new markets and to attract new types of visitor. That is not to say that the Forest of Bowland AONB's tourism and local produce are set to compete with national brands in the open market, but rather that the market is right for places like Bowland to celebrate and capitalize on their authenticity and sense of place.

Developing the Toolkit

Developing a Sense of Place Toolkit was identified as a high priority action in the *Forest of Bowland Strategy for Sustainable Tourism* published in January 2005. This was because it was seen as an essential starting point for many other areas of work: awareness raising, marketing and networking for example.

We recognized that the AONB Unit had limited resources to promote the area to visitors, and therefore we aimed to raise the area's profile via partners' marketing. In order to do this, consistent messages and information needed to be relayed to partners and we decided a toolkit would enable us to do this (see Figure 3.3). The toolkit was therefore produced primarily for tourism businesses in the Forest of Bowland AONB in order to help them utilize their own sense of place and local knowledge, which would then benefit their own business as well as the area as a whole.

The project started with a wide consultation exercise with over 300 people who lived and worked in the area. This was facilitated by Pathways, a Manchester based consultancy. The idea was to collect people's personal 'sense of place' through their thoughts, information and ideas about Forest of Bowland AONB and to stimulate conversation which could be recorded. A 'sense of place bag' was put together containing a map (to help locate the area and identify its extent), sample photographs to stimulate conversation, and a number of question cards. These included questions and prompts for discussion on topics such as:

- Folklore, myths and legends
- Must-see places in Bowland
- Where to see wildlife
- Favourite paddling and picnic sites
- Recommended walks and cycle rides
- Memories of the area
- Changes and traditions in farming
- Best ice cream, cheese and meat

The bag was taken to several 'focus groups' locally including a young mothers group, a pensioners' luncheon club, a parish council meeting, a volunteer ranger group and so on. The exercise culminated at the annual Hodder Valley Show in September, when people (including local families and visitors) were encouraged to fill in the cards in return for a free ice cream!

There was an amazing response and a huge amount of information was gathered and the outcome of this process was the development of the interpretive themes mentioned earlier.

The toolkit, then, takes a three stage approach for each of these themes. Firstly we supply factual information, for example a short history of the area, or a description of notable wildlife habitats and species found in the Forest of Bowland AONB – this is provided as downloadable text on a CD (and later, on the website)

Figure 3.3   Cover of the Sense of Place Toolkit
Source: Forest of Bowland AONB

together with images which are copyright-free so any business can use them free of charge. This gives the reader some background information to help them learn more about the Forest of Bowland AONB, or inspires them to research a particular issue which interests them in their own area. The text and images can be simply cut and pasted to the individual's website or publicity material.

Secondly, relevant information generated by the consultation is provided, for example a list of people's favourite ice cream, places to see wild deer, or recollections of living on a farm. This is left open for the business to contribute their own thoughts and ideas, or it can be passed on to visitors as recommended ways of discovering the Forest of Bowland AONB. Finally, action points and resources are provided to help the business create their own sense of place. This might for example suggest creating a 'Bowland breakfast' for guests by putting together local eggs, bacon and sausages to give visitors a taste of the area; or they could plan self-guided walks from the door to encourage visitors to explore the neighbouring countryside; or put together a local calendar of events so visitors can plan to visit when the heather is out, or when there are bird-watching events on. The Resources section then signposts the business to websites, publications and organizations which can help further.

The toolkit also features examples of best practice where businesses are already utilising their assets to develop a strong sense of place. These include a pub which sources and promotes local producers (see Figure 3.4); a farm where visitors can find out more about farming practice through the year, and get up close to lambs, piglets and wild birds (see Figure 3.5); a community celebrating the annual blooming of the bluebells; and a holiday cottage which celebrates its past owners, residents and traditions. These examples were included to inspire others, and we now regularly see examples of businesses utilizing a sense of place to help promote their product – opportunities to go hare-watching; wildife blogs on the website; an information room stocked with walks leaflets, maps and guides; drawings and photographs of the house and owners through the generations.

The toolkit was designed by Countryscape (a UK consultancy specializing in both sense of place and design communications) and published using the thematic colours, high quality images and print: and the new 'Bowland brand'. Quality is an important element of our work, and we also encourage partners to work to similar high standards so that visitors will associate the area with good information, good service and a visit that exceeds expectations. The purpose behind the toolkit approach, and the subsequent training, is to say to the businesses – you are the local experts, you have the knowledge (or the potential to gather that knowledge) and your visitors are curious about the place – so share it, talk to them, celebrate the sense of place and encourage them to come back! Repeat visitors are valuable to tourism, so establishing a relationship with people and encouraging their love of the place makes good business sense.

**Figure 3.4   A chef at the Three Fishes Inn at Mitton, which is well known
for its high quality food made using local produce**
Source: Forest of Bowland AONB

**Figure 3.5   Cobble Hey Farm, run by Edwina and David Miller,
             encourages visitors to find out more about their working farm
             and the wildlife around them**
Source: Forest of Bowland AONB

Training and Subsequent Developments

The toolkit was initially distributed to over 300 known tourism enterprises in and around the Forest of Bowland AONB, as well as to partner organizations such as district tourism officers, and the Lancashire and Blackpool Tourist Board. We later decided to supplement the publication with a training day which would bring the issues to life, and provide an opportunity to discuss the issue and to support businesses in developing their own sense of place.

We initially ran four training days in 2006/7 and attracted over 50 people from businesses, partner organizations and Tourist Information Centres. Subsequent events in 2007 and 2008 attracted a further 40 people including volunteer rangers and a number of artists involved in running the Bowland Arts Festival. The AONB Unit recognizes the value of these events, and is committed to running one or two a year for the foreseeable future in order to cater for newcomers to the tourism sector in the area, and as a refresher course for some of our more committed partners.

The training day, like the toolkit, aimed to convey the facts about the Forest of Bowland AONB and its special qualities, but also to encourage people to explore their own sense of place, to identify their special interests (be it natural history, food or walking for example) and to develop an action plan to deliver this to their customers. The training was informal and interactive and consisted of a pub quiz, fieldwork identifying locally distinctive features in the local area using a Polaroid camera, and brainstorming to develop ideas. The training also included a short lecture on marketing and sense of place by Paul Mahony, illustrating how major marketing campaigns often use this theme of locality in their selling of products from whisky, through to fashion and food (see above).

Satisfaction with the courses was high, and incidentally sparked off several long-lasting friendships and business relationships. The courses also led us to recognize that our tourism businesses, often sole traders or a family run concern, were keen to talk to one another, discuss trends and opportunities and to collaborate. Our first Sense of Place training course, held at Chipping Village Hall, was popular with over 20 businesses attending, and it overran considerably due to this desire to network. This, combined with other events and needs, led us to set up the Forest of Bowland Sustainable Tourism Network, a group of over 100 tourism enterprises which has now become a not-for-profit company, run by the member businesses themselves: Bowland Experience Ltd.

Our hope was that trainees would return to their business and implement their ideas: some did, but not all. So we moved into Phase 2 and began to offer one-to-one support in web management, whereby an experienced web designer worked with individual businesses to integrate their ideas into their own websites. This led to the development of some fantastic work on partners' websites about the Forest of Bowland AONB and about their own areas too – wildlife blogs, photographs of local views, downloadable walking routes and heritage information.

We are now considering how to incorporate recent work on a Landscape Character Assessment for the Forest of Bowland AONB into the Sense of Place work. Landscape Characterization is a system of landscape classification which relies on identifying generic features, but also locally distinctive qualities (which can include enclosure patterns, building styles or even sounds and smells). In the Forest of Bowland AONB 14 local character types have been identified and mapped including areas such as moorland plateau, lowland farmland with parkland and reservoirs and forestry. These types are then broken down into locally distinctive areas which may differ from neighbouring areas due to underlying geology, human influences or tree cover for example. Utilizing the sense of place work could help develop this classification by adding locally held information on history or farming practices, or illustrating it with photography, video or other artwork.

Impacts of our Work and Conclusions

The Forest of Bowland AONB has pioneered a business-led approach to sustainable tourism for protected areas in the UK, and our sense of place work has been an important ingredient in the success of this work. This is because rather than take a lead on marketing the area ourselves (as an AONB organization) we supported the businesses to market themselves as a part of the Forest of Bowland

AONB. This was partly because we did not have the resources for such a massive marketing campaign, but mainly because we wanted to work in partnership with tourism enterprises in a way which could be continued into the future, as part of our commitment to sustainability.

Our sense of place work has been applauded by regional and national partners including Natural England, Natural Tourism North West and the Lancashire and Blackpool Tourist Board. Other protected areas, and tourist destinations are seeking to replicate the approach – something we see as our biggest accolade – as a sense of place is something we can all share and celebrate.

Locally the impact of our work is beginning to pay off. We increasingly see our business partners promoting themselves as a part of the Forest of Bowland AONB, and also using their own sense of place to develop a distinct identity and appeal for their visitors. Our regular summer visitor survey is seeing a rise in the number of respondents who recognize what an AONB is, and the fact that they are visiting one. It is also identifying a rising trend in the number of people who are coming to visit the Forest of Bowland rather than a specific attraction, village or activity – we are finally becoming a destination in our own right.

The wider and longer term impacts of this collaborative approach to sustainable tourism are also visible. Tourism businesses have developed a real commitment to the Forest of Bowland AONB – and we are committed to them:

- individuals have become trustees of a new not-for-profit company and charity to support future developments;
- over 100 members in the business network have signed up to a commitment to sustainability;
- local collaborations between tourist enterprises are developing; and
- a continuing drive for quality and recognition as seen in the number of award wining businesses in the Forest of Bowland AONB.

We are continually inspired by this enthusiasm and innovation, and our commitment to our partnership and the sense of place work which it is based upon, will hopefully continue to inspire others into the future.

The overall outcome of re-branding Bowland is that this once hidden landscape is now becoming an established destination in its own right. To the potential visitor it is no longer portrayed as simply 'the countryside', a placeless area through which people would only journey on their way to the Lake District or Yorkshire Dales. Its unique qualities, stories and characteristics have been brought to the fore and given personality by the people who understand and value them. And perhaps more importantly, there is a genuine sense of shared ownership of this achievement. It is a success won without the need for major investment or high profile publicity, but instead driven by the AONB Unit and its many stakeholders working together to interpret and nurture the area's identity 'from the bottom up' – resulting in a brand, a sense of place, that truly belongs to the area and reflects all that is local and distinctly Bowland.

References

Forest of Bowland AONB, *Business Enterprise Survey; Key Findings – Changes between 2004–2009* [Online]. Available at www.forestofbowland.com/files/uploads/pdfs/strategies/Business%20Enterprise%20Survey%20key%20findings%2004-09.pdf [accessed: 15 April 2010].

Forest of Bowland Visitor and Enterprise Surveys 2005–10 [Online]. Available at www.forestofbowland.com.

Relph, E. 1976. *Place and Placelessness*. London: Pion.

Research Box, in association with Land Use Consultants and Minter, R. 2009. *Experiencing Landscapes: Capturing the Cultural Services and Experiential Qualities of Landscape* [Online: Landscape Character Network, Natural England commissioned report NECR024]. Available at http://landscapecharacter.org.uk/resources/reports/experiencing-landscape [accessed: 15 April 2010].

Simms, A., Kjell, P. and Potts, R. 2005. *Clone Town Britain: The Survey Results on the Bland State of the Nation*. New Economics Foundation. Available at www.neweconomics.org/sites/neweconomics.org/files/Clone_Town_Britain_1.pdf [accessed: 15 April 2010].

United Nations World Commission on Environment and Development 1987. *Our Common Future*. Oxford: Oxford University Press.

Visit Wales 2005. *How Do I Develop a Sense of Place?* [Online: toolkit to download]. Available from http://new.wales.gov.uk/topics/tourism/toolkits/senseofplace/?lang=en [accessed: 15 April 2010].

Chapter 4

Memory and the Value of Place in Estonia

Gurly Vedru

Introduction: People and Landscape

People are connected with places and landscape. Relations with places are different among different people, as is the sense of place and their understanding of it. The most intimate relations are often with places where one has lived all one's life or over a longer period of time. Personal connections and meanings arise over the course of time, and places become connected with us and with our ancestors in this way, adding temporal depth to these relations. This chapter is concerned with the relations between people and their environment, between people and the landscape they inhabit. It raises the questions: What is important for a local community? Is this familiar landscape valuable, and if so, then what precisely gives It that value?

The connections between people and their surrounding landscape has been different in different times and places. In the past, as amongst present-day traditional communities, relations were probably more intimate and deeper than they are today. The landscape was often considered to be animated, and something people communicated and interacted with (Taçon 2000: 50). Although landscape is a physical entity, it is also socially constructed in the minds of people (Children and Nash 1997: 1), or a matter of perception as the European Landscape Convention puts it. So its importance was/is not only economical, but also cognitive.

People have always sought to understand and make sense of their surroundings, whether it be the landscape as a whole or some of its elements. These explanations and reasons for searching landscape for stories and meaning probably derive from a deeper sense of place and landscape experience and from an accumulation of personal connections with it, no matter if it comes directly from one's own experience or is passed down through generations. Especially in the latter case, an oral tradition, connected with some places in the landscape can play an important part in developing this relationship with place (e.g. Taçon 2000: 50). Landscape is one of few things that connects people over time – the same landscape that is inhabited today was also inhabited millennia ago. Of course, past landscape differed from that of the present, but prominent landscape features will often persist. Changes have taken place: once forested areas may be open now; bodies of water have disappeared or turned into bogs; rivers may have changed their courses; but some features will persist (Vedru 2007a: 37).

Every place in the landscape has the potential to be meaningful for its inhabitants; it has its meaning and story, some kind of importance, hierarchy, biography and *genius loci*. The significance of places, however, can be judged to be different: some places will be deemed more important than others. These values can be culturally defined, or they can be locally held, or they can be both. Landscape bears multi-layered meanings and symbols; it is laden with knowledge and with memory.

This chapter poses two specific questions that together address the persistence of place and perceptions of value over time. First, what landscape features have been important for people in the past? And second, how do people in the present evaluate such places and their wider surroundings? Answering these questions I shall analyse two small areas, rich in archaeological sites and a well-preserved natural environment (Map 4.1). My first case study concentrates on the villages of Rebala protected area, its landscape and people and on their attitude towards their surroundings both in the past and the present. My second case study presents the problems and future plans of the people living in Tõdva-Kajamaa-Lokuti area.

Map 4.1   Map showing location of places mentioned in text. 1 – Tallinn; 2 – Rebala; 3 – Tõdva-Kajamaa-Lokuti area

Source: Drawing by Kersti Siitan

Rebala Past: People and the Landscape

Rebala is a Heritage Reserve in the vicinity of Tallinn, the capital of Estonia. It coveres 70 km² and comprises 350 recorded archaeological sites most of which are stone-cist graves and cup-marked stones of the Bronze and Pre-Roman Iron Ages. At present it is the only protected area in Estonia which seeks the preservation of a prehistoric agricultural landscape and its characteristic elements: archaeological sites, historical villages and buildings.

Most of the protected area is located on the North Estonian Plateau that rises about 40m above sea level. A small part is on the North Estonian Coastal Plain. These two zones are separated by an area of North Estonian Glint. The latter is one of the most remarkable natural phenomena in North Estonia, being a relatively steep terrace rising to a height of 35m.

There are two rivers in the study area: the River Jägala at the Eastern edge and the River Jõelähtme that flows through Jõelähtme village. In the karst region of Kostivere the River Jõelähtme goes underground and runs there for 2.5km. It surfaces in the southern part of Jõelähtme village. The large karst region of Kostivere is situated southeast from the centre of present Jõelähtme village. In this region both *loos* (in Swedish *alvars* and elsewhere loess, a thin humus-rich soil) and thicker moraine soils can be found. The *loo* areas were the first that were used for agriculture in North Estonia. A few damp areas are located in the lands of the Rebala village and they probably mark the locations of what were previously bogs. The main feature of the landscape is an open *loo* area, where views open up for several kilometres. *Loo* areas are situated in the coastal area of Northern and Western Estonia. In North Estonia *loo* areas are near the glint on the Limestone Plateau. The areas north and west from the Rebala and Võerdla villages were effected by the phosphorite mining, carried out in the 1980s, that has left deep openwork pits surrounded by high soil mounds. In other places the landscape remains free of these influences.

Most of the villages of the Rebala area are old and have retained their plan form. In some cases even the original farm buildings survive, situated in the same places as in the eighteenth century (Troska 2007).

Preparation for establishing the protected area of Rebala began in the 1970s, when new archaeological sites were located and those previously recorded were checked. The idea of establishing the protected area was that of an archaeologist, Vello Lõugas, who also carried out most of the fieldwork here at that time. At the same time as these archaeological studies, the chemistry factory of Maardu, wishing to extend its phosphorite mining, started to exert pressure on the local area. If these plans had been executed, several older villages and archaeological sites might have been destroyed. In 1979, a protected area of local importance was created in Rebala, its aim being to preserve historic landscape. A protected area of wider (republican) interest was established in 1987 (Kraut 2007; Pärtel 2007: 7). At present Rebala is the only area of heritage protection outside Estonian towns.

My personal interest in these areas started in 2005 when I carried out an archaeological survey in the village. Exploring a strange and unfamiliar place revealed new places that presented questions and caused me to search for answers in the local landscape. These searches, walks and discussions with local people inspired me to study this topic more thoroughly (c.f. Vedru 2007a). So, in 2006 I carried out a total inventory of all archaeological sites in the Rebala protected area (Vedru 2007b).

As we have seen, the area is rich in archaeological sites and monuments. The oldest archaeological sites date to the Mesolithic period (ca 7000–4000 BC). These are settlement sites, found near rivers. In fields today one can find pieces of worked quartz that indicate settlement sites of the Mesolithic (Vedru 2007c: 41). But the Stone Age people did not change their surroundings in ways that would last. The only exception was probably cutting off early growth in the preliminary forests in present *loo* areas (c.f. Lang 2000: 104; Vedru 2002: 108–9; 2007a: 44).

Changes to people's worldview and beliefs, and through that also their use of the landscape, found ultimate expression in the Bronze Age and pre-Roman Iron Ages. These changes left visible marks also on the study area where a large number of stone-cist graves and cup-marked stones are known. Both of these monument

Figure 4.1   A stone-cist grave in Rebala
Source: Photo by G. Vedru

classes form groups, although sometimes they are found in isolation. Although the graves and stones are often in similar situations, it is not always so. All stone graves are situated on dry land and possibly also on higher ground (Figure 4.1), but some of the cup-marked stones are located on the edges of damp areas or even in the middle of them (Figure 4.2). These latter places can be considered liminal. These are places at the edge, where ordinary meets different, and they are places perceived as special. These were the places where alvar met the bog, where high limestone plateau ended suddenly, and where the river that runs underground comes back into view. These places were often used differently, mostly for burying the dead. Cup-marked stones are found in these locales. Somewhere between these liminal places were the settlement sites and ordinary everyday lived landscapes. These landscapes had their own meanings and were experienced and perceived in different ways. All these places together formed part of peoples' worldview, their self-determination and understanding of the world (Vedru 2007a).

The cup-marked stones of the villages of Rebala are rather big and clearly visible on the landscape. Some are located a short distance from each other and these groupings have clear visual boundaries. The ground surface here is gently undulating, and although the differences in height are marginal, the views from

Figure 4.2   A cup-marked stone situated in a small bog in the village of Loo in Rebala protected area

Source: Photo by G. Vedru

most of the stones are wide and far-reaching. One can begin to suggest that some important places were changed through human activities while others remained unchanged. Some important features of the study area were ridges and lower terraces, glint, karst and probably also bogs. A large number of stone graves have been built on or near them: cup-marked stones can also be found in these places. Wide views are possible from graves and stones and several natural objects are visible in the distances. People moved between these places, experiencing and interpreting their surroundings. A number of graves and cup-marked stones are located in places that can be considered liminal.

There are some stone graves from the later part of Estonian prehistory; a number of pit grave cemeteries and several villages that were established in the Viking Age. The same villages still exist today with a continuous history of occupation of 1,000 years and more.

But these are not the only valued items or places in Rebala. There is a medieval stone chapel, a church and a large number of old farm buildings, dating to the eighteenth–nineteenth centuries. Stone fences, old roads, and windmills help to create a typical landscape of North Estonia. Rebala was one of the first areas in Estonia where phosphorite was mined in the 1920s. Old factories, built near the glint remain.

During the Soviet period the area of Rebala was mainly used for agriculture. Some buildings, used by collective farms still exist, although most of them are no longer used. Part of the landscape was destroyed by phosphorite mining which now took place in the open and over extensive areas.

So the landscape of Rebala area was created over several thousand years by the inhabitants who cut down the forests, built their houses and buried their dead. The agriculture has been carried out since the Pre-Roman Iron Age and possibly even earlier (Lang et al. 2001: 35 and references). As the same areas were used for agriculture also in later centuries, the landscape that exists today is open. Extensive views open up between the groups of stone graves, between stone-graves and natural objects (Vedru 2007a). Open views are also available to the church and chapel that once were considered landmarks (Vedru 2007b; Figure 4.3). In that way Rebala can be considered a well-preserved landscape where old meets new, but with an emphasis on the past. Different layers of landscape can be separated: prehistoric, medieval, historical and the present, all of them with distinctive elements that together inform the area's historic character.

Rebala Present – People and the Landscape

Rebala is close to Tallinn, the capital of Estonia. Economic growth at the end of the 1990s and beginning of the 2000s brought with it a rise in building activity. During the last few years a large number of new houses have been built in the area surrounding Tallinn. This rise in real-estate has now reached Rebala and people are parcelling up their farmland and selling it for development. The value

Figure 4.3   View to the chapel of Saha from the north-west
Source: Photo by G. Vedru

of this landscape is now more economic and less emotional, begging the important question of balance: how can the historic value of Rebala be preserved while at the same time allowing it to be developed for the future. Solving this problem is being achieved through spatial planning which has highlighted areas with officially-held high values for natural, archaeological and historic places and also areas for future real-estate development.

Within this spatial planning framework, a questionnaire survey was conducted amongst local people. Meetings with local people were arranged several times both in single villages and across the whole area. The positions of local people were investigated in order to map people's interests and opinions. The questionnaire was handed-out in meetings and was also available on the Internet. It showed how a large number of people consider their surrounding landscape as being low in value and with no need for protection. Several questions were asked, considering the open alvar areas of Rebala, about possible new houses and new people who would move into the villages of Rebala. These questions were asked of almost 1,700 people, though only 105 answered. It is a small number of people, yet probably the more active and caring part of the local community.

It was found that less than half of the respondents valued their home village, and that almost half valued the protected area of Rebala. One question asked whether the alvar area, the main feature of local nature, was worth preserving. Only a quarter of respondents thought it was. Paradoxically, the villagers from the alvar area did not value it highly, and inhabitants of other areas valued them more.

Half of the respondents thought that villages should maintain their original groundplan: that is, if the village originally had houses sited haphazardly, then new houses should also be placed in this way. Half of the respondents also thought that new houses should be traditional, not differing much from the old ones, meaning that traditional materials should be used, that the buildings should not be very high and/or have modern architecture. Most of the people liked the idea of new people moving into the area.

So it seems that most members of the local community want to change their surroundings, sell their land in parcels and build new houses. The landscape in itself does not have high emotional value for the local community. It is more a landscape laden with potential economic value.

Tõdva-Kajamaa-Lokuti: Past, Present and Future

Another case study comes from the area of Tõdva-Kajamaa-Lokuti (Figure 4.4). These villages are located ca 20 km to the south from Tallinn. The area measures approximately 17 km² and there are 15 archaeological sites registered including three settlement sites, dating probably to the last centuries of Estonian prehistory (i.e. eleventh–thirteenth centuries AD), one metallurgic complex, three stone graves and five cup-marked stones. Also an 'offering–spring', 'offering-stone' and a battlefield, originating probably from the Middle Ages and/or Early Modern period are known. As an addition, local people have supposed the existence of a holy grove in the village of Tõdva. A little further from the core area of these three villages is a probable refuge place and a pit-grave cemetery. The latter may also belong to the Middle Ages and/or to the Modern period.

The oldest archaeological sites – cup-marked stones and stone graves – are dated to the Bronze- and Pre-Roman Iron Ages, i.e. from the time interval 1500 BC to 50 AD. Settlement sites are recorded from all three villages meaning that all of them have some prehistoric background.

Tõdva and its neighbouring villages Kajamaa and Lokuti are set within a primarily agricultural landscape, and separated from other areas comprising forests, wetlands and bogs. The only large body of water is Vääna River that flows on the northern edge of the area. At present the landscape is open and the terrain undulating. The present villages consist mostly of dispersed houses and farms. Archaeological sites of the area are located close to each other; sometimes their distance is only a few hundred metres.

Compared with Rebala, only a few archaeological investigations have been carried out in Tõdva-Kajamaa area. Rescue excavations were conducted in one settlement site in Tõdva village revealing it to be an unusual metallurgic complex

**Figure 4.4   Tõdva-Kajamaa-Lokuti area. Typical landscape in the village
of Kajamaa**
Source: Photo by G. Vedru

dating to the eleventh–twelfth centuries AD. It is an important archaeological site,
largely because it is the only known metallurgic complex that dates to the last
centuries of the prehistoric period on the eastern shore of the Baltic (Kiudsoo and
Kallis 2008: 178). As an addition, archaeological investigations have been carried
out in the settlement site of Kajamaa (Vedru 2008). All the archaeological objects
have been visited and described by a history student of Tallinn University, Reigo
Andok (Andok 2009).

As it is not a protected area, the isolated archaeological objects are protected
only with their closest surroundings, mostly encompassing a 50m margin. Since
the Tõdva-Kajamaa-Lokuti area is also situated not far from Tallinn, this area is
also under pressure from real-estate developers. For example, one area chosen for
development is a small area between recorded archaeological sites. If everything
goes as planned, there will be 95 small houses built side-by-side between and
in the surroundings of cup-marked stones and settlement sites of Lokuti village.
The overall number of planned houses is still much bigger, and can be counted in
several hundreds, but they remain further from known archaeological sites. If all
these houses are built, their existence will change the character of this landscape
permanently. Unlike the people of Rebala, the local inhabitants of Tõdva-Kajamaa-

Lokuti area oppose this building programme, presuming that such a development will 'spoil' the landscape, have a detrimental effect on archaeological sites and destroy the present living environment. Local people have started actively looking for solutions, ways to maintain their present landscapes and to avoid the disturbance of archaeological sites not yet discovered. That there may still be surprises was made evident in the case of Tõdva settlement site, that happened to be a large metallurgical complex. In the search for help, a society comprising citizens of the three villages has been formed. The aim of this society is to look for possibilities to preserve the present state. For investigating their possibilities, locals have turned to the National Heritage Board proposing a protected area that would cover these three villages and their closest surroundings. They have written two letters, one of which summarizes archaeological information from the area and also all information they have about old roads, fords, mills and other places known through oral tradition (Ots 2008a). In the second letter, the values of the area are also expressed and it contains a discussion of the possible solutions to this dilemma (Ots 2008b). In this second letter they have also formulated their aims, that are as follows: first, to study and to protect the old villages; second, to create signage for interesting archaeological sites, and to map and interpret them; third, to make an agreement between local inhabitants considering how to balance the needs of development and heritage protection (Ots 2008b). Their dilemma and their attempts to find a solution have been published in the newspapers (*Maaleht* 1 October 2009).

At present, there is no solution to these problems, but the first steps have been taken towards finding one. Economic depression has caused a standstill in the real-estate market and so it has offered at least temporary easing of the situation. But that is most probably only a pause and change will continue in the near future. At the same time it is a suitable moment to create a society of local villages, though how it will act and what effects it will have, remain to be seen.

Peoples and Attitudes

Rebala and Tõdva-Kajamaa-Lokuti are very different areas, both in their general appearance and in people's attitudes. People of Rebala are mainly indigenous, whose ancestors have lived here for centuries; in Tõdva-Kajamaa-Lokuti most inhabitants are newcomers. The latter have valued the privacy given by the environment and they want to keep it that way. The protection of antiquities will support their efforts. The interest of the settlers of Rebala is more economic and their attitude to heritage protection is that it could prevent the real-estate development that they favour. For this reason they are not supportive of their cultural heritage. This does not mean that all people of Rebala support (unlimited) development and that all inhabitants of Tõdva-Kajamaa-Lokuti are against it. But the consensus is obvious in each case.

Similar work has been carried out in the national park of Karula, southern Estonia. Interviewing local people gave evidence that those people who have been living in Karula for a long time tended to view land as a resource and newcomers valued it as a peaceful and quiet environment for living (Jääts 2010).

One interpretation of these case studies is that people can get used to their surroundings to the extent that they will no longer recognize its value, even though it exists. They seem to forget that places have meanings that may be lost if one considers only material interests. Sometimes these values are more visible to the outsiders, who may have a fresh view of it. Sometimes one only notices things when they have gone.

Conclusion

It is not yet known what will happen in Rebala and Tõdva-Kajamaa-Lokuti, but one thing does seem certain: the landscape will change, at least to some degree. It is important but also difficult to find a balance between the developers and those seeking its protection. The spatial planning of Rebala will hopefully provide a useful framework for helping to keep the most valued places intact and allowing development in areas considered to have lower value. The future of the Tõdva-Kajamaa-Lokuti area remains still unknown.

Places are different. They have their own identity and history. People shape them and value them, giving meaning to the landscape and to places within it. Cultural landscape is thus a creation of the people who inhabit it. If a place within the landscape has a strong identity, then it also has a strong sense of place.

Acknowledgement

I would like to express my gratitude to Maarja Zingel from Space and Landscape for letting me use the results of their questionnaire. The present research was supported by the Estonian Science Foundation (grant No 6998).

References

Andok, R. 2009. Tõdva, Lokuti ja Kajamaa eelrooma rauaaja muistised maastikul. Unpublished manuscript.
Children, G. and Nash, G. 1997. Establishing a Discourse: The Language of Landscape. In G. Nash (ed.), *Semiotics of Landscape, Archaeology of Mind. BAR International Series, 661.* Oxford: Archaeopress, 1–4.
Jääts, L. 2010. Maaelu ideoloogiad ja praktikad. Karula rahvuspargi ja sealsete elanike näitel. – http://dspace.utlib.ee/dspace/bitstream/10062/14661/1/jaats liisi.pdf.

Kiudsoo, M. and Kallis, I. 2007. Metallurgic Complex in Tõdva Village, Harjumaa. *AVE 2006*, 175–82.

Kraut, A. 2007. Rebala kaitseala eellugu, in *Maa mäletab ... Valitud artiklid aastatest 1977–2007, Pühendatud Vello Lõugase mälestusele ja kaitseala 20. aastapäevale*, edited by M. Pärtel and M. Kusma. Jõelähtme, 12–39.

Lang, V. 2000. Keskusest ääremaaks. Viljelusmajandusliku asustuse kujunemine ja areng Vihasoo-Palmse piirkonnas Virumaal. *(MT, 7.)* Tallinn, 9–369.

Lang, V., Laneman, M., Ilves, K. and Kalman, J. 2001. Fossil Fields and Stone-cist Graves of Rebala Revisited. *AVE 2002*, 34–47.

Ots, T. 2008a. Tõdva kandi muinsusväärtused. Unpublished manuscript.

Ots, T. 2008b. Kas muinsuskaitseseltside aeg on tagasi? Unpublished manuscript.

Pärtel, M. 2007. Maa mäletab ... in *Maa mäletab ... Valitud artiklid aastatest 1977–2007. Pühendatud Vello Lõugase mälestusele ja kaitseala 20. aastapäevale*, edited by M. Pärtel and M. Kusma. Jõelähtme, 5–11.

Taçon, P.S.C. 2000. Identifying Ancient Sacred Landscapes in Australia: From Physical to Social. In W. Ashmore and A.B. Knapp (eds), *Archaeologies of Landscape. Contemporary Perspectives*. Oxford: Blackwell, 33–57.

Troska, G. 2007. Rebala muinsuskaitseala põliskülade ajaloost, in *Maa mäletab ... Valitud artiklid aastatest 1977–2007. Pühendatud Vello Lõugase mälestusele ja kaitseala 20. aastapäevale*, edited by M. Pärtel and M. Kusma. Jõelähtme, 81–112.

Vedru, G. 2002. Maastik, aeg ja inimesed, in *Keskus – tagamaa – ääreala. Uurimusi asustushierarhia ja võimukeskuste kujunemisest Eestis*, edited by V. Lang. Tallinn: Tartu, 101–22.

Vedru, G. 2007a. Experiencing the Landscape. *Estonian Journal of Archaeology*, 11(1), 36–58.

Vedru, G. 2007b. Rebala muinsuskaitseala arheoloogiamälestiste inspekteerimise aruanne. Unpublished manuscript.

Vedru, G. 2007c. Asustusjäljed Rebala maastikel, in *Maa mäletab ... Valitud artiklid aastatest 1977–2007. Pühendatud Vello Lõugase mälestusele ja kaitseala 20. aastapäevale,* edited by M. Pärtel and M. Kusma. Jõelähtme, 40–8.

Vedru, G. 2008. Aruanne arheoloogilistest uuringutest Kajamaa asulakohal. Unpublished manuscript.

Chapter 5

Being Accounted For: Qualitative Data Analysis in Assessing 'Place' and 'Value'

Stephen Townend and Ken Whittaker

Introduction

In the UK we can draw on 'values-based' historic environment practices that have emerged from recent advances in conservation planning. However, applications in the planning and environmental regulatory sphere are likely to be more demanding and nuanced than is the case where management of the historic environment dimension is the principal concern. In particular such practices are now more likely to be subject to greater testing, through environmental and planning law, as a consequence of recent policy changes.

As consultants working within the planning system, our role is to give consistent and transparent advice to customers, be they private or public sector, on the ways in which proposed development schemes will or will not affect the historic environment at a given location. Until now, what the 'historic environment' consists of has been relatively straightforward, comprising monuments, sites, buildings and various forms of historic areas. Assessment has also conformed to well established scales of importance and effect, which draws on statutory measures and designations, as well as national, regional and local policy provision. We know our 'data' and we know our 'methodology'.

But, with the publication of *Conservation Principles* by English Heritage in 2008, which reflects the *European Landscape Convention* (Council of Europe 2000), that data may be set to change beyond all recognition and adequate assessment methodologies have yet to be established.

Implications of Proposed Policy Changes

Article 1 of the *European Landscape Convention* states that '"Landscape" means an area, as perceived by people ...'. This means that any place or area does not exist solely as a physical entity, but exists also in as much as it is understood by anybody who engages with it.

In response *Conservation Principles* ascribes different types of heritage values, comprising evidential, historical, aesthetic and communal aspects (English Heritage 2008: paragraph 5) which constitute different approaches to the perception of landscape and 'represent a public interest regardless of ownership' (English Heritage 2008: paragraph 1.4) (Figure 5.1). It is the sum of these values as they relate to the historic fabric that define heritage significance, as referred to in the recently published Planning Policy Statement 5: *Planning for the Historic Environment* (CLG 2010a) and supporting guidance (CLG, DCMS and English Heritage 2010).

Given our involvement with planning and environmental regulation, we have a very specific viewpoint on the perception of place and the role of a value-based system for the definition and assessment of significance. This can be summarized under the following four headings:

1. We would wish to ensure that decision-makers are confident that the subjective aspects of perception, as defined in the *Convention*, interpreted in *Conservation Principles* and applied to UK planning policy, can be subject to a clear means of assessment. Any process needs to be readily understood, capable of being applied consistently and able to deliver the means to balance and weigh issues in a transparent, fair and reasonable manner.
2. That processes can be developed that are acceptable to the public interest.
3. Practices that are widely supported by practitioners can be commercially applied.
4. As a profession we should be able to exploit this opportunity, through the use of standard data capture systems, akin to the model of Historic Environment Records (HERs), in order to more systematically analyse how the historic environment is perceived than is presently possible. It is important that this is achievable both at and beyond the site scale as a means of informing future public policy, research and management practice.

Significance

The UK government has recently issued Planning Policy Statement 5: *Planning for the Historic Environment* (CLG 2010a), which introduces a unified approach to the management of the historic environment in England. The policies and principles set out in PPS 5 apply to the consideration of the historic environment in relation to the other heritage-related consent regimes for which English planning authorities are responsible. Management of change within the historic environment is set within a wider context of sustainable development, taking account of climate change, recognizing the potential value of the historic environment to place-making, regeneration, quality of design, minimizing waste and sustainable land use, as well as acknowledging the contribution of the historic environment to knowledge and understanding of the past. In order to manage change within the historic environment, PPS 5 sets out a significance-based approach to planning decisions, requiring sufficient evidence of the assessment of significance and appropriate measures to mitigate adverse effects on heritage assets and their setting.

Representations to consultative exercises prior to the adoption of PPS5 highlighted a number of concerns widely held across professional and voluntary sector conservation and planning interests. Whilst the concept of significance can be grasped, its application is not unanimously welcomed and there is confusion regarding its practical application (CLG 2010b). These concerns relate to both the application of significance and the ability to respond to community interests in a way that ensures these interests are properly engaged in the assessment process. Whilst subsequent published policy appears to offer a degree of reassurance (CLG 2010a), it is apparent that major institutions representing different aspects of the historic environment profession, including town planners, archaeologists, conservation officers and heritage volunteers, have been concerned that policy advances lack a sufficiently robust framework.

Accounting for perception in historic environment assessments therefore raises a considerable number of challenges both conceptual and practical. The conceptual recognition of *place* and *value* are foremost among these challenges. We suggest one approach that could resolve this from which we put forward procedures that we believe, can begin to address the concerns raised.

Place

The notion of *place* is clearly crucial, at least as it relates to the assessment of a set of associated values. 'Sense of place' implies experience of, immersion in, or at the very least an involvement with some locale or other. So, with the European Landscape Convention in mind, this paper puts forward a notion of place as a useable analytic unit which, rather than attempting to analyse a 'place' in and of itself, looks instead towards those that have an involvement with it and upon whom it may ultimately be understood to rely for its existence.

This understanding of place implies two key notions that require unpicking. First, a place is not akin to a box that contains things and within which things happen. Second, a place, in the sense intended, does not exist independently of those who have an involvement with it. So there are no physical boundaries, objects, structures or buildings that are necessarily implicated; rather a place is borne of an interpreted engagement with time, stories, associations, people, buildings, structures, objects, 'natural' features etc. and expressed as an understanding. A place may have an historic dimension, but this is not primarily a collection of attributes (listed buildings, sub-surface archaeology etc.) although it may manifest these; rather the historic dimension of a place is the totality of involvements which are related to both the idea of history or the past and the 'things on the ground'.

The 'historic environment' is not a Thing in and of itself, rather a conceptual entity that may or may not always reference physical objects. The implication is that historic environment 'places' are not necessarily (by which we mean that *it cannot be otherwise*) related to a set of features, sites or buildings – although they may be; rather that such 'places' are constructed through people's engagements with and understandings of them. Fundamentally therefore, *places do not exist outside the understandings of those who engage with them.*

This definition of place is well established in much recent archaeological research and in anthropology and is broadly acknowledged in *Conservation Principles*. As a result of the European Landscape Convention, the implication for Environmental Impact Assessment is that places so constructed should be subject to assessment. This is a substantive move away from current approaches as it means that it is not the 'place' *per se*, as a physical entity or the things in it that are subject to effects, but the *understandings* of a place as historically 'valued' in such-and-such a way as to represent a significance that may be effected by a development proposal. So in Environmental Impact Assessment language, the principal *receptors* are not the physical facets of an 'historic environment place', neither are they the people who engage with a place, although both need to be considered; rather it is the *understandings* that constitute the significance of such places that 'receive' the effect.

Value

Conservation Principles presents the notion of the value of places and defines how they might be understood. Similar categories and definitions have been forwarded by the Countryside Agency (CoAg) and Scottish Natural Heritage (SNH).

Debate on the meaning and definition of specific values has been and will continue to be long-running, but from the outset we should be clear that caution must be exercised in making presumptions about their importance for specific groups and communities. We should also recognize that the social factors that determine relationships with the historic environment may be complex and in may cases, no doubt, contradictory. And again there is evidence that practitioners have yet to reach a settled view on how values can inform practical conservation. As noted by the Institute for Archaeologists (IfA), Institute of Historic Building Conservation (IHBC) and the Association of Local Government Archaeological Officers (ALGAO, UK) (2008: 3.2), '[d]ebate continues about how best to characterize these sets of values, how they relate to each other, and what role they should play in decision-making'.

Nevertheless, to summarize place and value as implied by the European Landscape Convention, a place may be understood as a conceptual entity that is constituted through understandings that articulate the values associated with it as held by those who have involvements in it. These values are multi-faceted and include among them understandings of association, identity, remembrance,

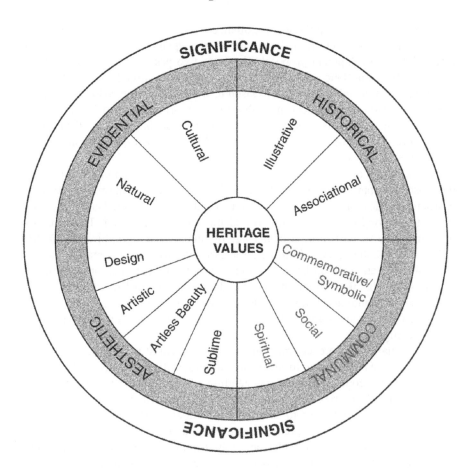

Figure 5.1   **Wheel diagram from second stage consultation on Conservation Principles (English Heritage 2008) illustrating the values that may be related to the historic environment**

coherence, community, sanctity, forgetting and many others. This chapter takes the position that places 'matter' and it is not *why* they matter that is important, rather *the ways in which* they matter that are of greatest relevance as it is these that fundamentally 'create' a place.

An Approach to Assessment

The sorts of values we are being asked to consider cannot be independently defined by a detached professional observer; they can only be articulated by those who have an involvement with a place. The role of the historic environment practitioner is to draw out these contextualized understandings of the value of an historic environment and to make an informed judgement on how they may be affected by a development proposal. There are two principal ways in which we might go about this. One would be to collect statistically valid data on a place. For instance, χ people value place γ for reason α, through a structured questionnaire. The other would be to collect more discursive information on a 'community's' understanding, accounts, discussions, visual representations etc. through an unstructured or semi-structured approach such as an open questionnaire, focus group or oral history.

We do not favour the former – quantitative – approach for a number of reasons, the most important of which is statistical relevance and bias. In order to be valid, a structured questionnaire would have to be randomly distributed and those most likely to respond are those with a vested interest of some kind. They are also most likely to be part of a particular age or income demographic. The DCMS *Taking Part* (2007) survey indicates, for example, that approximately 70% of all adults had attended a historic environment site at least once or twice in the previous year, but the highest rates of attendance were recorded within white and higher socio-economic groups. In addition, developments can galvanize small, motivated and vocal groups of detractors or supporters. The potential for bias is, therefore, high.

Contextuality and predetermination are also potential problems. The limited answers that are necessary to formulate meaningful statistical analyses do not lend themselves to a contextualized understanding of the heritage values of a place which may not always be clearly acknowledged by those who hold them. Structured questionnaires – which have to be brief or potential informants will not engage with them – run the risk of predetermining results and of imposing 'important' criteria through inclusion or exclusion. Only a limited number of questions may be asked and they may not address those aspects that a community itself finds significant; the questions themselves may unwittingly lead an informant in a particular direction and people may end up feeling excluded rather than involved in the process.

It is for reasons such as these that this chapter proposes a *qualitative* approach to values assessment. Such an approach aims to draw out the contextualized understandings relating to the value of a place as articulated by those who hold them.

Broadly speaking, this means gathering accounts of a place through some form of consultation process and analysing these accounts for recurrent themes and expressions of value. Although widely used in other sectors such as social research, education and nursing (see Seidman 1991 for example), this process generates a very different form of 'data' and requires a very different kind of analysis to that with which the regulatory arm of the heritage sector in the UK is familiar.

What is Our 'Data'?

Qualitative source data are potentially any discursive accounts or expressive representations that may range from the written and spoken word through to photographs, film and a wide range of other media, directed towards a 'case' (Miles and Huberman 1994), which for the consultancy sector is a proposed development scheme. Primary data (text, photographs, oral accounts and so on) may be gathered through a number of means, the most fruitful of which, in the context of development, may be some form of public 'open house' consultation whereby people are given a forum for discussing their understandings of an historic place.

Similar exercises have been undertaken as part of local community capacity building initiatives, based on heritage interpretation, which in South Wales, for example, have been supported by HERIAN (Heritage in Action), a partnership of 13 South Wales local authorities, national Statutory bodies, voluntary bodies and the private sector who promote the heritage of industrial South Wales for the benefit of both local people and visitors to the region. These 'Big Map' exercises, where local community participants contribute knowledge to a large scale map of a given locale on which they record specific elements of personal or otherwise assumed heritage or historic interest to give a portrayal of people and place, have been used to capture both qualitative and quantitative data, relying primarily on paper and photographic resources.

Beyond the UK and Europe similar mapping exercises have been used to engage Australian aboriginal groups (for example Hart 2001 and Harrison this volume), where values are clearly central to traditional relationships with landscape. These exercises might be usefully developed to include audio/visual records with associated commentary. There is also the opportunity to expand access through the use of Information and Communications Technologies, utilizing the web to engage a wider range of groups, including those with a special interest.

Various initiatives within the academic sector might also be worth examining, such as the joint Stanford Archaeology Center/Department of Archaeology at Göteborg University project *Co-creating Cultural Heritage*, which builds on earlier experiments in what Michael Shanks describes as 'deep mapping' (Shanks 2008), an approach that shares some outward similarities with the local Big Map interpretation plan exercises. Whatever the method, the aim would be to bring together stories, accounts, assertions, opinions and so on that are directed towards a place.

This would potentially provide historic environment practitioners with a form of context specific data through which to consider the constitution of a place; not a list of receptors, but a collection of discursive commentaries and representations. However, what such practitioners want, ultimately, is a list of receptors, so how do we deal with these 'data' to get at what we need in order to come to a meaningful assessment of the values associated with a place?

How Do we Deal with our Data?

Returning to the discussion of understandings as receptors, we need to draw out those understandings and give them short meaningful labels. This process is called 'coding' and there are a number of software packages that are specifically designed to facilitate this. In thinking through the issues discussed here ATLAS.ti, published by Scientific Software in Berlin was used, so it is elements of this software environment that will frame the following discussion.

Coding is a technically simple process but it can be conceptually very challenging. The first step is to create 'quotations' which are simply highlighted bits of text (or audio recordings, videos or photographs) that seem relevant; as in the example below where Norham Castle, Norham (Northumberland) is described as having a 'mystical aura', (Figure 5.2) or the photograph of a memorial in Berwick-upon-Tweed (Figure 5.3) where the boxes pick-out apparently relevant

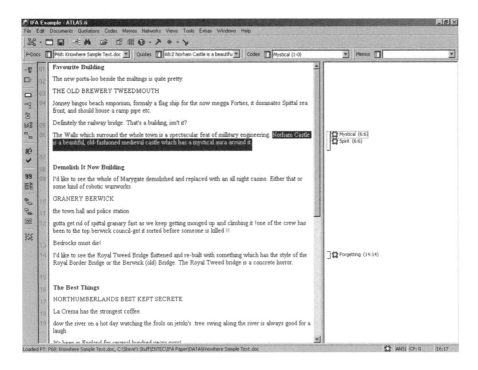

Figure 5.2   Textual account of Norham Castle showing an example of simple conceptual coding of passages in ATLAS.ti

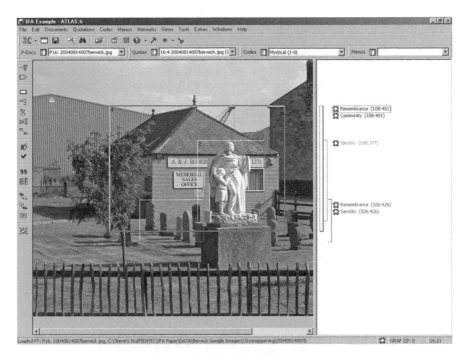

Figure 5.3 **Photograph of a Memorial at Berwick-upon-Tweed showing an example of simple conceptual coding of an image in ATLAS.ti**

bits of the scene represented. Next to these primary data documents is a pane that contains symbols and words. The bracket-like character shows which line(s) or area(s) is covered, the yellow symbol shows that this is a 'code' and the name (the conceptual label given to this bit of information) follows. In the following brackets is the *groundedness*: *degree of density*. These are analytical tools that, broadly speaking, help one to understand how 'real' one's chosen coding is likely to be.

One of the most challenging aspects of Qualitative Data Analysis (QDA) is choosing codes and there are a number of ways to do this. In these examples, we have picked a few that occur in *Conservation Principles* and added *mystical* and *forgetting* of our own. But passages (although not sections of non-textual documents) can be coded *in-vivo* which means that words are lifted directly from the quoted passage as can be seen in the following example.

In the Norham Castle text we could have selected 'mystical aura' which would have made a good code and represents the most direct way of drawing out articulated understandings from a piece of text.

The process of coding essentially gives us a second tier of data that is more amenable to analysis. For example, by looking at the number of times that a given code occurs, we can get some idea of the relative significance of the value of understanding that it represents. At a further level of analysis, codes can be

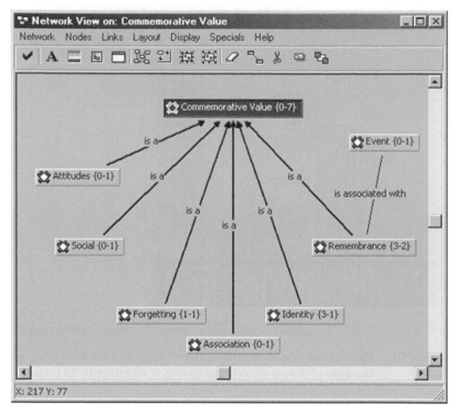

**Figure 5.4   Network Diagram from ATLAS.ti illustrating the connectivity
                  between concepts drawn out through coding**

connected together and given relationships that indicate the level of connectivity
between conceptual understandings leading to a more nuanced understanding of
the values associated with a place. This is shown in Figure 5.4 – an analytical and
display tool within ATLAS.ti – which shows a simple set of relations based on
Grounded Theory.

There is a great deal more that can be done with ATLAS.ti and other QDA
software packages, and it should be borne in mind that the examples are a highly
simplified use of Grounded Theory, which is only one of many approaches.
Nevertheless, a very basic analysis such as this can, through the degrees of
groundedness and density, begin to give a strong indication of the relative
significance of particular values for a given place.

Conclusions: What Does it all Mean?

It is possible that a values based approach advanced in *Conservation Principles* can embrace the democratizing principles of the European Landscape Convention. Traditionally management of the historic environment has relied on authority grounded in knowledge and was largely divorced from the associated communities. It is now proposed that the process should apply a greater degree of self-reflection and awareness of social context. However, there is good reason to question whether there has been adequate preparation to realize the opportunities this should afford local communities.

There is a split between the detached Cartesian perspective of an archaeologist operating in a professional capacity and the embedded lived experiences and memories that shape values important to the wider public. Clearly it would be wrong for practitioners to attempt to reconcile this division by drawing solely on their own apparently 'intuitive' or 'emotional' views based on a notion of a broadly common cultural experience. However, we have not established appropriate means of broadening engagement and currently lack the conceptual tools to analyse and reconcile multi-perspective understandings of heritage values. This potentially leaves decision-makers exposed and the planning-based system of heritage protection vulnerable. Yet, as has been briefly described, both the academic and voluntary sectors are engaged in activities that the regulatory sector could draw on.

With this in mind we have briefly put forward both a conceptual framework and a toolkit for approaching values, their interrelationships and introducing transparency in the analytical process for all participants in the decision making process.

The reason for offering this chapter is to focus attention on the need for conceptual and methodological responses and to advance constructive debate on the implementation of changes in UK policy and guidance. These policy changes themselves are to be welcomed as an important development in meeting the public interest. However, we consider that a more robust procedural framework is required if this ambition is to be realized. This framework should be applicable across all forms of development and address the key issue of assessing values as a means of defining the significance of place. Without such a framework it is difficult to see how policies that seek to further enhance and democratize planning decisions that affect the historic environment can be properly applied, as the evidence required to support an informed decision cannot be provided according to agreed standards of reliability. Therefore if recent policy changes are to be a firm response to the UK's ratification of the European Landscape Convention, there is still a need for corresponding procedures that will realize the 'significant scope to enhance the contribution the historic environment can make to the sustainable development and sustainable communities agenda, and to the emerging "place-shaping" agenda for local government' ('Foreword' in *Peer Review of English Heritage: Summary of Findings and Recommendations* June/July 2006).

References

CLG 2009. *Planning Policy Statement 15: Planning for the Historic Environment* (Consultation Draft). London : CLG.

CLG 2010a. *Planning Policy Statement 5: Planning for the Historic Environment.* London: TSO.

CLG 2010b. *Consultation on a New Planning Policy Statement 15: Planning for the Historic Environment. Summary of Responses.* Available at http://www.communities.gov.uk/documents/planningandbuilding/pdf/1514220.pdf [accessed: 9 May 2010].

CLG, DCMS and English Heritage 2010. *PPS5 Planning for the Historic Environment: Historic Environment Planning Practice Guide.* English Heritage: online.

Council of Europe 2000. *European Landscape Convention.* Florence: Council of Europe. Available at http://www.coe.int/t/dg4/cultureheritage/heritage/Landscape/default_en.asp.

DCMS 2007. *Taking Part: England's Survey of Culture, Leisure and Sport Annual Data 2005/06.* London: HMSO.

English Heritage 2008. *Conservation Principles Policies and Guidance for the Sustainable Management of the Historic Environment.* London: English Heritage.

English Heritage Living Draft – 24 July 2009. *PPS Planning for the Historic Environment: Historic Environment Planning Practice Guide.* English Heritage: online.

Hart, V. 2001. *Mapping Aboriginality. Investigating Queensland's Cultural Landscapes: Contested Terrains. Setting the Theoretical Scene.* Queensland University of Technology, Brisbane.

Heritage Link 2009. *Heritage Link Response to Draft PPS15: Planning for the Historic Environment and Draft Planning Practice Guide* [Online]. Available at http://www.heritagelink.org.uk/wp/wp-content/uploads/2009/10/Heritage-Link-response-to-PPS5-and-Practice-Guide2.pdf [accessed: 14 May 2010].

Historic Towns Forum 2009. *Draft PPS5 – Consultation Questions on Which Views are Being Sought:Response from the Historic Towns Forum* [Online]. Available at http://www.historictownsforum.org/PPS5_response [accessed: 14 May 2010].

IfA 2009. *IFA Response to PPS5 Consultation* [Online: Institute for Archaeologists] Available at http://www.archaeologists.net/modules/icontent/inPages/docs/consultations/PPS.pdf [accessed: 14 May 2010].

IfA, IHBC and ALGAO (UK) 2008. *Standards and Guidance for Stewardship of the Historic Environment.* Reading: IfA [Online]. Available at http://guide.opendns.com/track/click.php?q=Standards+and+Guidance+for+Stewardship+of+the+Historic+Environment&curl=http%3A%2F%2Fwww.archaeologists.net%2Fmodules%2Ficontent%2FinPages%2Fdocs%2Fcode

s%2FStewardship2008.pdf&search_grp=MAINRESULTS&search_pos=0 [accessed: 14 May 2010].

IHBC and RTPI 2009. *PPS 15 Planning for the Historic Environment, Incorporating Views from the Royal Town Planning Institute (RTPI) and the Institute of Historic Buildings Conservation (IHBC)* [Online]. Available at http://www.heritagelink.org.uk/wp/wp-content/uploads/2009/10/IHBC-and-RTPI-joint-response.pdf [accessed: 14 May 2010].

Miles, M.B and Huberman, A.M. 1994. *Qualitative Data Analysis: An Expanded Sourcebook.* Thousand Oaks: Sage.

National Trust 2009. Planning Policy Statement 15: Planning for the Historic Environment and Historic Environment Practice Guide, a Submission by the National Trust [Online]. Available at http://www.heritagelink.org.uk/wp/wp-content/uploads/2009/10/National-Trust.pdf [accessed: 14 May 2010].

Rescue 2009. PPS Planning for the Historic Environment Comments by Rescue: the British Archaeological Trust [Online]. Available at http://www.scribd.com/doc/21985885/PPS-15-Consultation-Rescue-Response [accessed: 14 May 2010].

Seidman, I.E. 1991. *Interviewing as Qualitative Research: A Guide for Researchers in Education and the Social Sciences.* New York: Teachers College Press.

Shanks, M. 2008. http://documents.stanford.edu/MichaelShanks/35 [accessed: 13 February 2009].

Appendix 1

Key Concepts

Place   and thereby an 'historic environment', exists primarily as a conceptual entity and only secondarily as a collection of physical ones.

Communal Values   relate to the constitution of an 'historic environment' and are manifest as understandings that reference a place.

Understandings   are the EIA receptors that may be affected by a development.

Primary Data   in a 'values-based' archaeology are the understandings that articulate those communal values that reference a place. They are manifested in accounts (both textual and photographic), opinions and actions.

Qualitative Analysis   for the purposes of this discussion, may be understood as a non-statistical approach to data that aims to illuminate the qualitative experience and interconnections of a place.

Chapter 6

'Counter-Mapping'
Heritage, Communities and Places in
Australia and the UK

Rodney Harrison

Introduction

Most people will be familiar with the experience of returning to a place known and
loved from one's past, only to find it altered, removed or demolished. The feelings
of loss which such an experience can engender are one poignant reminder of the
non-tangible or social attachments which we form to place, or what geographer Yi-
Fu Tuan (1977) and others (e.g. Feld and Basso 1996) have referred to as a 'sense
of place'. The social values of place, both at an individual and collective level,
have become an important new area of research in the field of archaeological and
cultural heritage management in Australia. This chapter summarizes recent and
emerging approaches to understanding and managing the social values of place in
Australia and reflects on their implications for interventions in heritage practices
in the UK. To do this, I will first provide a brief summary of the Australian heritage
system, before considering some background issues which have foregrounded
'sense of place' in Australian environmental and heritage planning. I will look
at some case studies developed to record social values of place from New South
Wales where I was previously employed in the NSW National Parks and Wildlife
Service Cultural Heritage Research Unit (now Department of Environment,
Climate Change and Water), focusing on the new techniques we developed to
'map attachment' and social values. Drawing particularly on case studies in
Indigenous cultural heritage management from New South Wales, I will outline
the ways in which archaeologists are increasingly engaged in a consideration
of both the tangible and intangible values of heritage sites, and discuss some of
the tools which have been developed to record and 'map' intangible values and
attachment to place in contemporary Indigenous, migrant and settler Australian
communities. Much of this work has taken the form of mapping and recording
alternate, 'hidden' or non-mainstream social geographies, and in the final part
of the chapter, I comment on the role of such 'counter-mapping' in giving voice
to politically marginal and subaltern understandings of the past and present and
consider an example of the use of such practices in the UK.

An Overview of the Australian Heritage System

Heritage in Australia is managed at all levels of Government, and different legislation governs heritage at the federal, state and local level. At the Federal level, Australia maintains a National Heritage List which contains natural and cultural heritage places of national significance. Places on the list are protected under the *Environment Protection and Biodiversity Conservation Act 1999*. In March 2008 there were 76 properties on the National Heritage List.

At the state level, each of the states maintains its own register of places of State heritage significance. In general, a distinction is made between the management and listing of 'Indigenous' and 'historic' heritage places, as well as 'natural' heritage places (see further discussion in Byrne 1996, Byrne et al. 2001, Harrison 2004). In NSW, for example, historic heritage places of State significance are listed on the *State Heritage Register* under the *Heritage Act, 1977 (amended 1998)*. Aboriginal archaeological sites and other sites of significance receive blanket protection under the *NSW National Parks and Wildlife Act 1974* and are listed on the NSW *Aboriginal Sites Register*.

At a local level, locally significant non-Indigenous heritage places in NSW may be listed in the heritage schedule of a local council's local environmental plan (LEP) or a regional environmental plan (REP) and receive protection under the *Environmental Planning and Assessment Act 1979*. Similar local heritage instruments operate in the other Australian states and territories.

'Sense of Place' in Australian Heritage Practice

Heritage practice in Australia is strongly influenced by the Burra Charter, or 'The Australia ICOMOS Charter for the conservation of places of cultural significance'. The Burra Charter was revised in 1999, partially due to a widespread feeling that the original version of the Charter, adopted in 1979, and its subsequent 1988 revision, had been too 'archaeological' and 'architectural' in focus. The 1999 revisions reflected a greater emphasis on social or intangible values and on 'places' rather than buildings and sites.

> The revisions broaden the understanding of what is cultural significance by recognising that significance may lie in more than just the fabric of a place. Thus significance 'is embodied in the place itself, its setting, use, associations, meanings, records, related places and related objects' (Article 1.2) ... The way the Charter deals with social value has been improved (through the recognition that significance may be embodied in use, associations and meanings); spiritual value has been included (Article 1.2); and the need to consult and involve people has been made clear (Articles 12 and 26.3) ... The Charter encourages the co-existence of cultural values, especially where they conflict (Article 13). (Australia ICOMOS Inc 2000: 22)

The Burra Charter has had a major influence on emerging trends in heritage practice in the UK, exemplified by the new English Heritage *Conservation Principles, Policies and Guidance* (English Heritage 2008) in which greater emphasis is given to 'communal value'.

The emphasis on the 'sense of place' and the experience of local distinctiveness of landscape in Australia grew out of the nexus of new approaches to social landscapes in archaeology and heritage and an interest amongst the broader Australian public in engaging with Indigenous social landscapes (e.g. Johnson 1994; Byrne 1996; Read 2000). This has emerged in the context of the Australian Government's slow, grudging (but now formal) recognition of the dislocating and disenfranchising effects of Government policies which placed many Aboriginal children with white families and led to the development of a 'stolen generation'. There is now a widespread acknowledgement of the need to heal the wounds of this policy through understanding and revitalizing the social connections between Aboriginal and non-Aboriginal Australians and their landscapes.

Case Studies

What follows is an outline of two case studies undertaken as part of the *Shared Landscapes* project over the period 2000–2004 (Harrison 2004). This project was focused on two concepts. First, to attempt to connect Indigenous and non-Indigenous heritage management and to look at the nexus between the two. Second, to examine ways in which landscape approaches to cultural heritage documentation might be operationalized in an Australian context. I will not dwell on the case studies as a full length monograph on the project was published in 2004 (Harrison 2004), and subsequent publications on aspects of the case studies have appeared elsewhere (Harrison 2003, 2005, 2010). I will briefly describe the case studies and then draw out the implications of the methodology for processes of cultural heritage management in more general terms.

Mustering Landscapes in Northern NSW

The first of these case studies sought to record and understand the heritage landscapes associated with cattle mustering in the Kunderang Ravines, an area now managed as a National Park and World Heritage Wilderness Area in northeastern New South Wales (Map 6.1). In addition to drawing on established archaeological, historical and architectural heritage recording techniques, the project employed a range of less conventional methods to map the 'landscape biographies' of both Indigenous and non-Indigenous former pastoral workers and their families, in the form of both mapped oral history, and of 'story-trekking' (after Green et al. 2003) along remembered narrative paths. Such an approach allowed a more embodied understanding of the landscapes of cattle mustering to emerge. By riding and walking along familiar pathways and mustering routes,

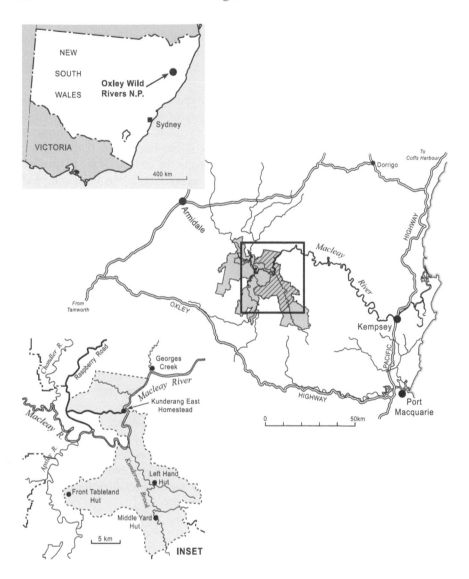

**Map 6.1   Map showing the location of East Kunderang and Oxley Wild
Rivers National Park in New South Wales, Australia**

pastoral workers and their kin created a familiar sense of being-in-the-landscape
(after Bender 2001), while simultaneously creating that landscape.

The oral histories of former pastoral station workers are rich with details of
mustering, riding and walking through the gorge country. A major theme of the oral
histories was mapping former mustering routes associated with Kunderang and

neighbouring pastoral stations, and discussing the appreciation of the landscape that people developed as a result of their passage through it. All mustering was done on horseback, and it was only in the 1950s and 1960s that motor vehicle access was made available at the homestead. Even during the 1990s, when Kunderang cattle were being mustered out of Oxley Wild Rivers and Werrikembe National Parks, it was done predominantly on horseback.

Interviewees were encouraged to make use of maps and aerial photographs at different scales to mark the locations of events and places to which they referred during oral history interviews. What many of the men and women drew was a series of lines that marked both physical tracks and pathways. The maps are rich with places which constitute landscapes of dwelling, working, walking and riding. These landscapes have a personal character, but also reflect wider shared notions of the landscapes in the pastoral industry. For the former pastoral workers and their descendents, the landscape of the Kunderang ravines is understood in profoundly different ways to those of the NSW National Parks and Wildlife Service who manage the national park. Recollections of the country emphasize particular kinds of places, such as clearings on the tablelands and river flats, the river itself, and the ever-important spurs – those escalators of the gorge-lands. The linearity and seasonality of movement between tableland and gorges forms a moving landscape, a construction within which people's memories can be articulated and made to speak in profoundly personal ways.

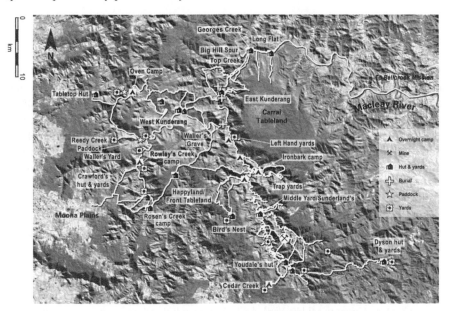

Figure 6.1   Mustering and travel routes associated with East Kunderang. Different informants' records appear here as a single line. Mustering huts, camps and yards are labelled

It is possible to represent all of these mustering tracks and pathways, along with the locations of huts and yards, on a map of the area now covered by Oxley Wild Rivers National Park (Figure 6.1). This map illustrates the patterns of pastoral land use in the Gorges at a landscape scale. A visual picture is conjured of generations of history lived in and through the landscape of the Gorges. This map also demonstrates that the area now managed as 'wilderness' has had a long history of thorough infiltration by Aboriginal people, cattle and pastoralists that has played a fundamental role in forming the landscape.

The relationship between work and people's understanding and appreciation of the landscape of the gorges is of critical importance. Jeff O'Keefe noted when describing a particular creek while he was mapping his landscape biography:

> Steep hard creek but amazingly enough we used to have very good luck because the sides were so steep, the cattle wouldn't climb out of them easily. They always used to sidle around the sides and then they'd come down again. Over the time we had a lot of success in it. Early in the piece we got every beast out of Blacks Camp. Some of the creeks – even Left Hand and Thread Needle – still have got a handful of cattle in them, but Blacks Camp – quite early in the piece we had every beast out of it. (interview, 2 March 2001)

The 'Short Cut'

For landscape philosopher Michel de Certeau, it is people's interlinked paths and pedestrian movements that form 'real systems whose existence in fact makes up a city' (1984: 97). The history of the city begins at ground level, with people's footsteps. In the Kunderang Gorges, it is not pedestrian movement but the movement of horses and riders along pathways, and cattle across their daily and seasonal 'beat', that constitutes the social face of the country. We can inscribe these movements and pathways as lines on maps, but to do so can mean that we miss the practices of starting and stopping, walking, crossing rivers, roping and throwing wild cattle, and incidents that occurred along the way (see Pearson and Shanks 2001: 148). De Certeau distinguishes tricks in the 'ways of doing' (1984: xviii), the ways in which people continually subvert the constraints of landscapes.

One such trick is the 'short cut', a frequent inclusion in people's oral histories which stress moving through space, constituting a focal point for the intersection between history, event, people and landscape (Figure 6.2):

> Yes it's not very far from the mouth of Thread Needle [Creek]. Well, actually you don't come out the mouth of Thread Needle with cattle, you short cut over a bank and cross to Middle Yard. And it's a steep little climb up and a steep little climb down and, in the dark, at night coming back with tired cattle, it was a great place where years ago they used to lose a lot of them. So we decided we'd take the portable yards, to a place where there was a bit of a track where they used to

cut posts years ago. We would just put them in to the portable yards without the hassle of losing them or widening the yard and we'd go back next morning, either take the 'Blitz' [truck] over and put them on, or drive them across next morning when we had plenty of time. (Maurice Goodwin interview, 1 March 2001)

Like the short cut, the detail of embodied landscape biographies can be lost in the broad stroke of the line on the page. The Kunderang narratives seem to support Gibson's 'theory of reversible occlusion', which describes the way in which the environment is known by humans along a path of observation of surfaces which move in and out of view in a particular order along a pathway or route of travel (1979: 198; see discussion in Ingold 2000: 238). These stories relive and recreate the landscape by recalling the routes along which it was experienced and known. The 'ways-of-doing' associated with mustering in the Kunderang ravines form part of the collective experience from which former pastoral workers constitute their sense of collective identity, and sense of place (e.g. chapters in Feld and Basso 1996). 'Places not only *are*; they *happen*' (Casey 1996: 13).

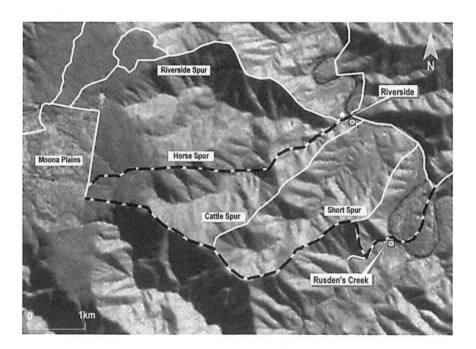

Figure 6.2   The 'short' and 'horse' spurs: short cuts used by the Crawfords to travel between Moona Plains, Rusden's Creek and Riverside

Dennawan

Turning to my second case study, the name 'Dennawan' describes a multiplicity of spatially concurrent places. It is principally associated with an unsupervised Aboriginal Reserve, gazetted in 1913 on the site of an earlier camp that had provided an Aboriginal labour force for surrounding sheep ranching properties.

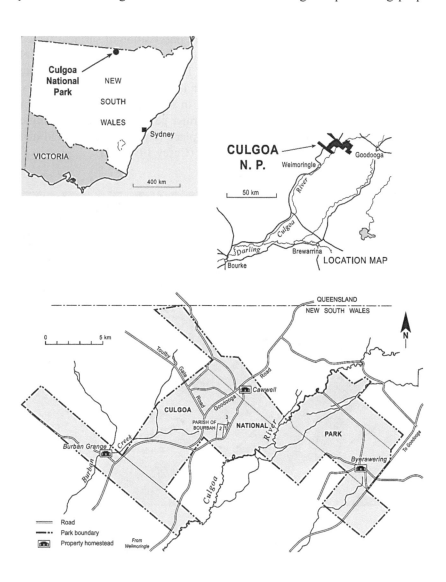

Map 6.2   **Map showing the location of Dennawan in New South Wales, Australia**

At the turn of the nineteenth century Dennawan was a bustling village; built at the junction of two travelling stock routes on the edge of the Western NSW pastoral frontier, it contained a hotel and an inn, a shop, a post office, a police station, and a resident Aboriginal population of several hundred people. Dennawan was also an Aborigines Inland Missionary outpost, where the fondly recalled missionary, Miss Ginger, taught children to read and write. Dennawan is an archaeological site on the edge of Culgoa National Park, a place visited and recalled in the present. Dennawan is a place from which Aboriginal people were removed in the 1940s – a symbol of the broader 'spatial story' (de Certeau 1984) associated with the NSW Aborigines Protection Board's concentration and segregation strategies of the late 1930s and 1940s (e.g. Goodall 1996). Dennawan is simultaneously all and more than any of these things. It is an entanglement of genealogies, a place where past, present, and future collapse (Map 6.2).

My first experiences at Dennawan occurred during a visit to the site with several local Aboriginal people who had either lived or had ancestors who lived at the site in the 1930s. The first thing that struck me was the way people interacted and articulated their relationship with the place in an 'archaeological' manner. By this, I mean that it involved interrogating, touching, and talking about the material traces of the former settlement. People also interacted with the place in a formal, performative way, which suggested it was more than a dead memorial to the past. Instead, Dennawan emerged through the course of my involvement in recording it not as a dead place but an active site for the contemporary creation of locality, community, and collective identity. While I was mapping the remains of the Reserve, I developed a parallel investigation into the significance of the remains to local Aboriginal people and the way in which that significance manifests itself during visits to the site.

Technical detail obtained from fine-grained differential GPS recording was integrated with anecdote and memory (Figure 6.3) in the mapping of the archaeological remains at Dennawan to produce a multivocal, textured representation of the site and to provide insights into a shared past. An artefact database linked to a hand-held computer and differential GPS was used to record all of the 8,000 artefacts and structural features at the site to a horizontal accuracy of ±4 centimetres. Digital audio recordings taken in the field were captured as a separate layer and integrated into the GIS. Oral accounts and archaeological mapping were combined to develop integrated data sources on which to base an interpretation of the archaeology of the former Reserve. The site recording was undertaken during multiple field trips over a period of approximately 18 months. This relatively protracted period of investigation was important for allowing the community the longer time frames they required to engage collaboratively and in a considered way with the research, and it was an important part of the project methodology.

For descendants of the Aboriginal people who used to live on the Dennawan Reserve, the dead often visit the living in dreams. Contemporary Muruwari people have a number of beliefs about relics and their relationship with ancestors that

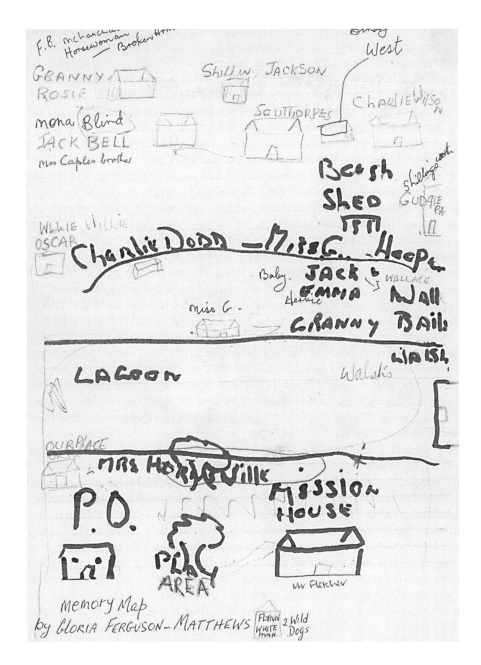

Figure 6.3   Gloria Matthew's 'memory map' of Dennawan

have contributed to the development of Dennawan as a place of pilgrimage. Physical contact of the body or skin with artefacts is considered a way of making a connection with the ancestral past. During site visits, Muruwari people like to rub artefacts such as those of flaked stone against their skin. Vera Nixon explained in an interview:

> When you're rubbing the stones over your skin you can get the feel of—you sort of get the feeling of the spirits coming into your skin somehow or another. I dunno, it's a strange feeling, but it's a good feeling. (Dennawan, 18 November 2001)

The belief that ancestors' spirits are associated with the objects they used during their lifetimes structures people's interactions with the remains of the former settlement. A trip to Dennawan, then, is much more than just an opportunity to learn about the past; it is an opportunity to make direct and intimate contact with it. Josie Byno said:

> When we go and visit the place and see the artefacts that they used to use and the fire there, the oven, we get very emotional. Not only that, there is a special feeling in the air that surrounds us. We can feel that spiritual feeling wherever we go, and we know that they are with us. (Dennawan, 18 November 2001)

While it is important for people to be able to touch and interact with the artefacts on site, it is considered dangerous to remove them. People who do this are tormented with bad dreams or sickness. In contrast, just being at the site is considered to make Muruwari people feel physically healthy. Arthur Hooper, now in his seventies, noted:

> Ever since I've been coming out here, doing a little bit of work for people, I've been feeling really great. I'm really happy to see the old place again. And my feelings—inside me it's a very glad feeling, I have no worries about anything else. No aches and pains, I just walk around the place for hours and hours without getting tired. (Dennawan, 18 November 2001)

The ability of the place to effect change on the body of Muruwari people is an important facet of the spirituality and significance of the former Dennawan Reserve. These corporeal influences are intimately tied to various spiritual associations with the former settlement, in particular, the slippage between post-1930 associations with Aborigines Inland Mission Christian missionaries and older, deeper associations with *wiyrigan* (medicine men) and *miraaku* and *miraga* (spirits). This slippage creates a certain denseness of experience that is felt by Muruwari people in the present when visiting the archaeological site, which they have increasingly done on a regular basis, especially over the past 10 to 20 years.

The archaeology of the former Dennawan Reserve has much information to contribute regarding the relatively hidden histories of Aboriginal pastoral labour camps in the nineteenth and twentieth centuries in Australia. However, the ruins of the former Reserve are much more than a source of information to local Muruwari people; they represent instead the focus for a program of shared, collective memorialization of the past. The artefacts that remain on the former Reserve are invested with intense emotional and spiritual power. They form the conduit for controlled interactions between the spirit and human worlds and between past and present. Instead of ceasing to exist after its abandonment, Dennawan continues to hold power and fascination for Muruwari people as a place where local traces and memories persist, challenging and actively assisting in the creation of the past and the present. It does this as much through the mutual involvement of people and objects, which both evoke and create collective memories, as through their absence or decay. Place and trace provide creative opportunities for citation, quotation, and montage (Pearson and Shanks 2001). For Muruwari people, Dennawan is past and future. Each trip to Dennawan represents an opportunity to excavate a 'place of buried memory' (Küchler 1999; Leslie 1999: 108) and to re-map Dennawan as a place in the present.

Other Case Studies from New South Wales

Similar work has been carried out by English (2000, 2001) and by Byrne and Nugent (2004) who undertook research with Aboriginal people from the mid-north coast of New South Wales, mapping pathways which connected mid-twentieth century settlements with coastal camping, fishing and picnic places. Byrne argues that:

> A key objective in all of this research was to bring to the broader public's attention the fact that, even in those parts of Australia that had been colonised by white settlers earliest and in greatest density, Aboriginal people had continued to maintain very extensive patterns of movement ... the mapping of such places has a 'counter' aspect to it in that it potentially unsettles the colonial mapping of resources which ... classified its usefulness in the framework of the colonial, not the Indigenous economy. (2008: 261, see also 2003)

Our concern with mapping people's sense of place and attachment to landscape developed within an environmental planning context in which heritage management is caught up in the business of large scale landscape planning. The map is fundamental to such approaches. However, the quote above suggests that such an approach should be considered as more than just a process of conceiving of ways of representing intangible values, but as a sort of 'intervention' in mainstream heritage practices. Byrne refers to this process as a form of 'counter-mapping'. I would like to explore the idea of counter-mapping and its role in heritage practice further.

Heritage and Counter-Mapping

Byrne (2008) describes the way in which Peluso (1995) developed the concept of counter-mapping to describe the maps produced by forest users in Indonesia to contest the State's maps which had been used in the past to exclude them from the use of forest resources. The term has subsequently come to describe the way in which maps are used to undermine power relations and challenge the dominant political and social geographies of power (Harris and Hazen 2006). Byrne (2008: 261) has noted that the need to be able to produce maps that make the spatial dimensions of indigenous cultures intelligible to their settler-(post)colonial neighbours has driven the adoption of counter-mapping practices and technologies (such as the development of collaborative GIS maps) by indigenous minorities in Australia and Southeast Asia. Similar initiatives have been reported amongst local communities in Kenya (Harrison and Hughes 2010), for example.

In the studies discussed above, the combined use of traditional practices of site recording with more intuitive practices of 'story-trekking', oral history mapping, and particularly the use of large scale aerial photography on which community collaborators were encouraged to draw not only led to a deeper understanding of the complex and multilayered attachments of participants in the studies with their landscapes, but also allowed us to deconstruct certain aspects of our own professional heritage practice. The interpretation and management of Oxley Wild Rivers National Park as a World Heritage wilderness area sought to stress the absence of human modification and history in the area. Instead, this process of mapping memories revealed the densely layered and complex histories of mustering in the gorge country. Similarly, at Dennawan, what we might have traditionally viewed as an abandoned archaeological site was re-animated as a place of contemporary pilgrimage and the active creation of a sense of community and identity in the present.

The processes of counter-mapping allow minority groups to challenge some of the 'taken for granteds' of heritage management, but also encourages people to celebrate their experiences of the everyday. This has the potential to draw more people into heritage practice, and to emphasize the active role of heritage in the production of identity, neighbourhood and community. Arjun Appadurai (1996: 180) has referred to these practices as technologies of localization:

> The building of houses, the organization of paths and passages, the making and remaking of fields and gardens, the mapping and negotiation of transhumance spaces and hunter-gatherer terrains is the incessant, often humdrum preoccupation of many small communities studied by anthropologists. The techniques for the spatial production of locality have been copiously documented. But they have not usually been viewed as instances of the production of locality, both as a general property of social life and as a particular valuation of that property. Broken down descriptively into technologies of house building, garden cultivation, and

the like, these material outcomes have been taken as ends in themselves rather than as moments in a general technology (and teleology) of localization.

Beginning to think of the practices of heritage and their role in the transformation of space to place is particularly ripe with potential for those people who might traditionally feel excluded from more 'mainstream' heritage places and practices. In the case of the Dennawan study, the Aboriginal participants had been removed from the Reserve in the 1940s and were displaced and effectively in diaspora (Harrison 2003). Documenting the social values of place through the study of everyday practices has the potential to put such people back on to the heritage map.

Thinking about counter-mapping raises ways in which local minority communities in the UK might engage with formal heritage practices, such as walking tours and bus tours, to develop alternate histories of place. I have recently interviewed Jay Brown who runs a Black Heritage walking tour of Brixton which is aimed at achieving this very thing (see Harrison 2010; see also Purbrick and Schofield 2009 for a study of the Brixton landscape). Her tour – around one of South London's notorious suburbs associated with the 1980s race riots – is aimed at changing the ways in which people think about the history of Brixton, and the place

Figure 6.4   Jay Brown in the Granville Arcade Market, Brixton, South London

of British African Caribbeans in British history and contemporary culture (Figure 6.4). When most people think of the heritage of London, they might think of the Houses of Parliament, Big Ben, Madame Tussaud's and the Tower of London. Jay's walking tour is unique in exposing tourists to a part of London which they might otherwise not experience due to fear or its apparent invisibility from the tourist landscape. But more importantly, Jay's tour puts an emphasis on Brixton's multicultural community, and celebrates its distinct British African Caribbean heritage. In doing so, it not only changes the way in which outsiders view Brixton, but also celebrates Brixton as a place and in doing so helps to build Brixton as a community.

Jay advertises her business on the internet, via postcards which she leaves for distribution with local businesses, and through word of mouth. Although Brixton is not a traditional destination for tourists or visitors to London, the area and the 'vibe' of its local African Caribbean community are central to her business. Jay notes,

> A lot of people are afraid to come to Brixton. They have heard about the riots, they think it could happen again and they have a misconception. By starting the tours, it's a key for people to see what Brixton has to offer ... If you think about the places they [tourists] have to go to [in London] it's pretty limited. It's fine if you want to see a palace, fine if you want to go on the Wheel [the London Eye], but I figured people needed something which was off the beaten track, like a hidden gem, and that's what Brixton is.

She sees her tour as targeted at those who wish to learn more about Black British culture and at independent travellers who want to be exposed to a different side of London life.

> There are a lot of independent travellers who want to come and do their own thing ... Lots of Black Americans come here and they want to see how [Black] people live in London, and not just travel around the West End ... This is a place I think people need to come to, to see what is going on in London. I'm opening the door for them to see different cultures. There are over 40 languages spoken in Lambeth alone. You can walk along the street and eat Japanese food, Portuguese food, Indian food ... if you come here you feel like you have escaped London and arrived somewhere else.

Jay's tour begins at Atlantic Road near the Dogstar pub, almost immediately moving on to Railton Road and the frontline of the 1981 riots. She takes this opportunity to discuss the riots and to try to help people to understand what prompted them.

> The tour starts here at Atlantic Road and straight ahead is the scene of the riots on Railton Road. So it's a nice start to the tour because we can walk up the street and look at the area and see the ways in which the area has regenerated since the

riots. And as we walk along I can explain to them about why the riots happened in 81 and 85, and the reason why is that a lot of the Jamaicans had free run of the area. They could have *shebeens*, which were late-night parties well into the night, and they could sell drugs. And they did this from the frontline, and the frontline was on Railton Road. And there came a time when the police were stopping and searching the young men, and anger built up in the community from that. But I think the main thing that set it off was that they felt that there was racism against them because 'stop and search' was mainly being levelled at young Black men. And it's not a nice thing, being stopped in the street and having your pockets turned out. And that was their way of saying they weren't going to take it anymore. And although that's not the way that things should be done, it was like a kettle boiling and at some point it had to go off.

Turning into Kellet Road, Jay highlights the Victorian terraced houses and tells of the community of artists and musicians who squatted in them in the 1960s, and the association of the area with Reggae musician Bob Marley. Walking past the Effra pub, Jay directs her tour towards St Matthew's Square and St Matthew's Church which date from the 1820s. Travelling along Brixton Hill and past the Tate Library, she recounts the association of the Tate family with sugar plantation slavery and the invention of the sugar cube. On Brixton Road she points out Windrush Square to discuss the connection with the first Jamaican community to arrive in London in 1948, and the Ritzy Cinema, the oldest functioning cinema in South London which opened in 1911 as the Electric Pavilion. From there she walks to the Brixton Academy, a major concert venue that opened in 1929, and past Bon Marché, the first department store in the UK when it was opened in 1877.

The tour ends at the Granville Arcade markets which opened in 1938 and form one of Europe's largest African Caribbean food markets, and nearby Electric Avenue, the first street in London to be lit by electricity, made famous by the song 'Electric Avenue' released in 1982 by Guyana-born British Reggae singer Eddy Grant.

Jay's tour is only one of a number of recent initiatives in Black British tourism and heritage in London. In 2003 Nana Ocran published *Experience Black London: A Visitor's Guide*, which included a series of Black heritage walking trails and a guide to Black heritage sites in London. Steve Martin, author of *Britain's Slave Trade* (1999), runs an open-top Black history bus tour of London's West End. In an interview with the BBC, he noted:

> We have a very narrow attitude of history … Shakespeare alluded to a multicultural London. But then we got an empire. And with that came a model of history that supported the idea of racial purity. What is most shocking is how little is taught in schools. Children know more about black American history than the heritage of people walking in their own streets. (Casciani 2003)

These initiatives can be thought of as interventions in official heritage practice in the sense that they use official heritage practices to draw attention to an overlooked aspect of heritage. Black British and British African Caribbean heritage has been largely ignored by official heritage listing and interpretation in London, although more has been done in the wake of the 2007 bicentennial celebration of the abolition of the British slave trade. These small-scale interventions all work towards changing the ways in which tourists and locals view the heritage of London and, by extension, the culture and community of London in the present. But more importantly, Jay's tour can be thought of as a work of heritage 'counter-mapping' in action. By walking the streets and speaking to visitors and locals she is actively engaged in a process of building new histories and new narrative trails through the landscape. Such initiatives have a strong connection not only with the ways in which visitors perceive the place, but also in contributing to the development of a new sense of place for the local community.

The other important aspect of Jay's walking tour is the way it commemorates and contextualizes the 1981 riots. It attempts to remove from Brixton a historical stain that has influenced tourists' and Londoners' perception of it as a place. Looking at the riots within the context of British African Caribbean music, food and culture might be seen as an attempt to balance the frequent negative press which presents Brixton and South London in general as a place associated with a complex cultural and ethnic mix that produces high rates of criminal activity. But perhaps more importantly, it reclaims and celebrates the riots as an important historical event that can be argued to have brought about positive social changes.

> When I explain to people they can empathise with the situation and understand why it happened, and then to take them to the scene it's a positive part of the tour. I emphasise the ways in which the riots set off a chain of events for change, and the ways the community has rehabilitated since then. Looking around you today you will see that Brixton is as trendy as Notting Hill. It's nice for people to come and see and … understand that out of such a bad thing came all of these good things.

By working with a well-known heritage tourist practice – the walking tour – Jay Brown has developed a heritage activity that has the potential to change perceptions of this infamous part of South London as well as the potential to create a sense of community and locality. This provides a clear example of Appadurai's ideas about the social work involved in the production of locality (1996) and its social relations. Jay Brown's walking tour literally counter-maps and (re)creates Brixton as a place by walking people along its streets and unveiling the connections between its community and the neighbourhood. In doing so, it reinforces Brixton as a multicultural community and the social relations that connect the community with the place.

Conclusions

The new interest in social value in Australian cultural heritage management has presented both heritage managers and communities with challenges in representing and integrating people's sense of place into conventional landscape planning processes. Nonetheless, various inroads into this process have occurred as a result of collaborative projects which have focused on mapping and 'counter-mapping' minority and hidden social geographies and people's attachments both to places from which they have been historically excluded, as well as the quotidian spaces in which they live. This process has the potential to give voice to politically marginal and subaltern understandings of the past, empowering them by drawing attention to them in the present. Formal heritage projects which seek to map counter-narratives in addition to local initiatives like Jay Brown's walking tour have the potential to reveal the complexity of landscape and people's sense of place, while celebrating the distinctiveness of everyday experience. Such intimate, everyday attachments of people to place are at the heart of contemporary approaches to heritage.

References

Appadurai, A. 1996. *Modernity at Large*. Minneapolis and New York: University of Minneapolis Press.

Australia ICOMOS Inc 2000. *The Burra Charter: The Australia ICOMOS Charter for Places of Cultural Significance 1999, with associated Guidelines and Code on the Ethics of Co-existence*. Burwood, VIC: Australia ICOMOS.

Bachelard, G. 1969. *The Poetics of Space*, trans Maria Jolas. Boston: Beacon Press.

Bender, B. 2001. Introduction. In B. Bender and M. Winer (eds), *Contested Landscapes: Movement, Exile and Place*. Oxford and New York: Berg, 1–18.

Benjamin, W. 1999. *Selected Writings*, edited by M. Jennings, M. Bullock, H. Eiland and G. Smith, trans R. Livingstone, vol 2. Cambridge: Belknap Press of Harvard University.

Byrne, D. 2003. Nervous Landscapes: Race and Space in Australia. *Journal of Social Archaeology* 3(2): 169–93.

Byrne, D. 2008. Counter-mapping: New South Wales and Southeast Asia. *Transforming Cultures eJournal* 3(1): 256–64.

Byrne, D., Brayshaw, H. and Ireland, T. 2001. *Social Significance: A Discussion Paper*. Hurstville: NSW National Parks and Wildlife Service.

Byrne, D. and Nugent, M. 2004. *Mapping Attachment: A Spatial Approach to Aboriginal Post-contact Heritage*. Sydney: Department of Environment and Conservation (NSW).

Casciani, D. 2003. Changing the Guard with Black Tourism, *BBC News Online* (16 July) [online]. Available at http://news.bbc.co.uk/2/hi/uk_news/magazine/3068973.stm [accessed: 5 March 2008].

Casey, E.S. 1996. How to Get From Space to Place in a Fairly Short Stretch of Time: Phenomenological Prolegomena. In S. Feld and K.H. Basso (eds), *Senses of Place*. Santa Fe: School of American Research Press, 13–51.

de Certeau, M. 1984. *The Practice of Everyday Life*. Berkeley: University of California Press.

English, A. 2000. 'An Emu in the Hole: Exploring the Link Between Biodiversity and Aboriginal Cultural Heritage in New South Wales, Australia', *Parks* 10(2): 13–25.

English, A. 2001. *The Sea and the Rock Give us a Feed: Mapping and Managing Gumbaingirr Wild Resource Use Places*. Hurstville: NSW National Parks and Wildlife Service.

English Heritage 2008. *Conservation Principles, Policies and Guidance*. London: English Heritage.

Feld, S. and Basso, K.H. 1996. *Sense of Place*. Santa Fe: School of American Research Press.

Gibson, J.J. 1979. *The Ecological Approach to Visual Perception*. Boston: Haughton Mifflin.

Goodall, H. 1996. *Invasion to Embassy Land in Aboriginal Politics in New South Wales, 1770–1972*. St. Leonards, NSW: Allen & Unwin/Black Books.

Green, L.F., Green, D.R. and Neves, E.G. 2003. Indigenous Knowledge and Archaeological Science: The Challenges of Public Archaeology in the Reserva Uaçá. *Journal of Social Archaeology* 3(3): 366–98.

Harris, L. and Hazen, H. 2006. Power of Maps: (Counter)-mapping for Conservation. *Acme International E-journal of Critical Geographies* 4(1): 99–130.

Harrison, R. 2003. The Archaeology of 'Lost Places': Ruin, Memory and the Heritage of the Aboriginal Diaspora in Australia. *Historic Environment* 17(1): 18–23.

Harrison, R. 2004. *Shared Landscapes: Archaeologies of Attachment and the Pastoral Industry in New South Wales*. Sydney: UNSW Press.

Harrison, R. 2005. 'It Will Always Be Set in Your Heart': Archaeology and Community Values at the Former Dennawan Reserve, Northwestern NSW, Australia. In N. Agnew and J. Bridgeland (eds), *Of the Past, For the Future: Integrating Archaeology and Conservation. Papers from the Fifth World Archaeological Congress (WAC 5)*. Los Angeles: Getty Conservation Institute, 94–101.

Harrison, R. 2010. Heritage as Social Action. In S. West (ed.), *Understanding Heritage in Practice*. Manchester and Milton Keynes: Manchester University Press and the Open University, 240–76.

Harrison, R. and Hughes, L. 2010. Heritage, Colonialism and Postcolonialism. In R. Harrison (ed.), *Understanding the Politics of Heritage*. Manchester and Milton Keynes: Manchester University Press and the Open University, 234–69.

Ingold, T. 2000. *The Perception of the Environment: Essays in Livelihood, Dwelling and Skill*. London and New York: Routledge.

Johnson, C. 1994. *What is Social Value? A Discussion Paper*. Canberra: Australian Gov. Printing Service.

Küchler, S. 1999. The Place of Memory. In A. Forty and S. Küchler (eds), *The Art of Forgetting*. Oxford and New York: Berg, 53–72.

Leslie, E. 1999. Souvenirs and Forgetting: Walter Benjamin's Memory-work. In M. Kwint, C. Breward and J. Aynsley (eds), *Material Memories: Design and Evocation*. Oxford: Berg, 107–22.

Martin, S.I. 1999. *Britain's Slave Trade*. London: Channel 4 Books.

Merleau-Ponty, M. 1962. *Phenomenology of Perception*. London: Routledge.

Ocran, N. 2003. *Experience Black London: A Visitor's Guide*. London: London Small Business Growth Initiatives.

Pearson, M. and Shanks, M. 2001. *Theatre/Archaeology*. London and New York: Routledge.

Peluso, N.L. 1995. Whose Woods are These? Counter-mapping Forest Territories in Kalimantan, Indonesia. *Antipode* 27(4), 383–40.

Purbrick, L. and Schofield, J. 2009. Brixton: Landscape of a Riot. *Landscapes* 10 (1), 1–20.

Read, P. 2000. *Belonging: Australians, Place and Aboriginal Ownership*. Melbourne: Cambridge University Press.

Tuan, Y. 1977. *Space and Place: The Perspective of Experience*. Minneapolis: University of Minnesota Press.

Chapter 7

Exploring Sense of Place: An Ethnography of the Cornish Mining World Heritage Site

Hilary Orange

Introduction

The Cornish Mining World Heritage Site provides an interesting example through which to examine local residents' understanding of 'sense of place'. The de-industrialization of Cornwall's tin and copper mining industry and the subsequent transition to an economy largely dependent on tourism, has inevitably brought many changes – from noise to quiet, from an emphasis on subterranean to the surface, and from physical exertion to a visual consideration of industrial ruins within a 'natural' setting. These changing perceptions have occurred within recent history and therefore, in large part, within living memory.

Retired miners and people who have direct associations with the mining industries live within the World Heritage Site alongside incomers who now make up a significant proportion of the local population. Indeed, the scale of the World Heritage Site is indicated by the fact that an estimated 81,535 residents live within the designated area (Cornwall County Council 2009).

During ethnographic research in Cornwall in 2008 and 2009 local residents within the World Heritage Site were asked what, if anything, the term 'sense of place' meant to them. The aims were to: a) gain better understanding of different public definitions of the term; b) test the relationship between definition and other demographic variables for statistical significance; c) consider descriptions of 'their' place; and d) inform comment on the appropriateness of the term and its uses within the archaeology/heritage sector.

Following a background which briefly contextualizes the research in terms of mining history, culture and economics, this chapter will discuss the themes and results that have emerged from this research. These include the different ways in which 'sense of place' was found to relate to concepts of place, cognition and belonging; the relationship between place of birth and definition; and how descriptions or 'senses' of Cornish mining places appear to be most commonly described in romantic and aspirational terms. In conclusion, the possible implications for the heritage management professions, in Cornwall and further afield, will be discussed.

Cornish Tin and Copper Mining – A Brief History

Cornwall is located in the far South-West of the British Isles. While Cornwall is currently administered as an English county its constitutional status is being questioned by some at a time when the UK is experiencing significant change through devolution. It is within the 'county's' central spine of granite that the tin and copper mining industry was located. The archaeological remains of the industry can be broadly dated from the seventeenth century ending, specifically, on 6 March 1998, when South Crofty, the last tin mine in Cornwall, closed (Thorpe et al. 2005: 21). At its peak, in the early nineteenth century, the industry supplied two-thirds of the world's copper (Rowe 1953: 128) and its scale and importance is evidenced through the extent of relict mining sites (including associated industries such as ports, harbours, foundries and arsenic production) which can be found across Cornwall's coastline, towns, moors and villages.

De-industrialization began in the 1860s and proved a long-drawn out and painful process (Brayshay 2006: 131, 142–3; Thorpe et al. 2005: 8, 11). Newly discovered metals and fast developing mines, notably in Chile, South Australia, Michigan and Cuba, caused copper prices to crash in 1866 (Schwartz 2008: 60). Thereafter tin became Cornwall's principal product, yet foreign competition and the difficulties of extracting tin ore at greater depth beneath the unprofitable copper lodes caused tin prices to fall in the 1870s. By the 1890s the industry was facing extinction and with only nine mines left in production thousands of Cornish miners emigrated to South Africa, Australia, Mexico and other foreign countries where mining appeared more lucrative (Payton 2004: 219–20). In the face of economic collapse two 'projects' sought to 're-invent' Cornwall. First, tourism began to be viewed as an economic panacea; by the late nineteenth century Cornwall had become associated with literary tourism, due to the works of Hardy, Swinburne and Tennyson being set in the county and in the early twentieth century the 'Cornish Riviera' poster campaign of the Great Western Railway ensured Cornwall's popularity as a tourist destination. Second, a Revivalist movement sought to create a 'new' post-industrial identity by looking back to the Celto-Catholic period through a revival of Celtic language and arts (Payton 1997: 28, 36).

Twentieth-century mining was dominated by the super-mines, such as Geevor and South Crofty, which were rich and large enough to survive fluctuations in metal prices. In the post-war period mining entered a phase of optimism following a programme of government funding and soaring tin prices (Brayshay 2006: 143; Schwartz 2008: 87). By 1951 a popular belief in a Celtic-Cornish identity was strong enough that a nationalist party, *Mebyon Kernow* (Sons of Cornwall) was formed (Deacon 2007: 209–10). Interventions to conserve and 'clean up' the relict mining sites began in the 1970s and 1980s with statutory reviews, moorland clearance subsidies, land reclamation schemes and works by local societies and The National Trust following acquisitions (Palmer and Neaverson 1998: 129; Schwartz 2008: 62, 120; Thorpe et al. 2005: 18). As spoil heaps were removed (sometimes viewed as eye-sores, sources of contamination, or removed for the re-processing of secondary minerals (Palmer 1993: 49–50)), mine-shafts capped and mine buildings repaired, the cultural value of Cornish mining heritage increased. In terms of the remaining 'live' mines the optimism of the post-war years proved to be a swan-song. Tin prices had been kept artificially high and when the International Tin Council collapsed on 24 October 1985 the price of tin halved overnight. Despite a desperate struggle for survival, including appeals to central government, the last surviving mines closed (Brayshay 2006: 143; Schwartz 2008: 87). In subsequent years several mine sites, including Geevor, have re-opened as heritage centres and museums.

Pride in Cornwall's former position as an industrial world leader remains embedded within a modern sense of Cornish identity. Indeed the flag of St Piran – the patron saint of tin miners which shows a white cross of tin ore against a black background of charcoal – is a recurring symbol throughout the county (Payton 2004: 285). Mining areas may still be places of commemoration (for what has been lost) but they have also now become places of leisure, heritage and tourism. In 2006, in recognition of the international significance of Cornish mining (for the period 1700 to 1914), 10 mining areas in Cornwall and West Devon were awarded World Heritage Status (Cornwall County Council 2009). Interestingly, as comparisons were being made to the Taj Mahal and the Pyramids, there was also increasing 'chatter' within the county about the possibility, with increases in metal prices and demand from China, of mining returning to Cornwall again. Most hopes rest on the current development works at South Crofty (Schwartz 2008: 74).

Exploring Sense of Place

To examine local senses of place more closely, four case study areas were selected, their diversity of form and location a deliberate attempt to achieve topographic and socio-economic variance (Map 7.1).

Case Study Areas

These case studies comprise of three mine sites (Botallack, St Agnes and Minions) whilst the town of Hayle is the fourth comparative site chosen for its urban, as opposed to rural, situation. The following short descriptions briefly describe the landscape setting and socio-economic activity within each of the four case study areas:

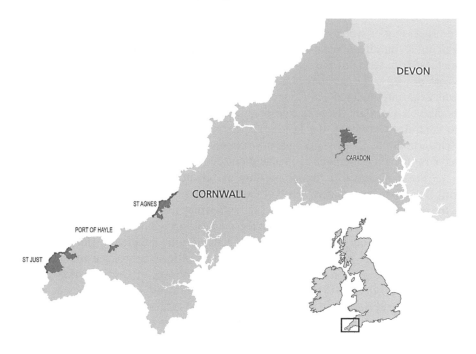

Map 7.1   The location of the World Heritage Site case study sites
Source: © Cornwall Council

Botallack (St Just Mining District)   Botallack, in the far West of Cornwall, is a small hamlet with a coastal strip of mining remains which lies in close proximity to Geevor Mine (closed in 1990) and the sizeable town of St Just-in-Penwith. Major mining activity started at Botallack in the nineteenth century, after which the site was reworked sporadically in the twentieth century before eventually sections of the cliff were acquired by The National Trust from the mid-1990s (The National Trust n/d). During its live period Botallack was renowned (Sharpe 1992: 99) for its submarine mines and the engine houses of Crowns Mine which perch precariously at the bottom of the cliffs (Figure 7.1). The site contains a range of mining features including chimney stacks, an arsenic labyrinth and twentieth-century dressing floors. Aside from National Trust members and coast-path circuiteers, most visitors to the site are local people walking their dogs or driving onto the dressing floors and sitting (often in their cars) to enjoy the sea-view.

**Figure 7.1   The perilously positioned engine houses of Crowns Mine,
                Botallack**

Source: Photo by Hilary Orange

Figure 7.2   Hayle Town
Source: Photo by Hilary Orange

Hayle (Hayle Mining District)   In the eighteenth and nineteenth centuries, the town of Hayle (situated around a large estuary on the North coast) was one of Cornwall's largest and most important ports which supplied local mines with coal and timber (Figure 7.2). Its character and layout is unusual due to its development in the early eighteenth century from two rival company towns. Distinctive styles emerged, with Copperhouse to the East (Cornish Copper Company) and Foundry to the West (Harvey & Co.'s iron foundry business) (Noall 1984). Since the post-war period the port and harbour area has become increasingly derelict and in recent years has been subject to a number of proposed development plans, the latest being Dutch property developer ING Real Estate's plans to revitalize the harbour area with a mixed-use scheme including over 1,000 new homes (ING 2009). Alongside prospective regeneration a number of community-led heritage projects (Harvey's Foundry Trust 2008; Hayle Area Plan Partnership 2009) have formed in recent years with the aim of working towards urban regeneration and raising awareness of Hayle's industrial heritage.

St Agnes (St Agnes Mining District)   The large village of St Agnes has two main foci – St Agnes Churchtown, which sits on a relatively flat terrace of land below the granite summit of St Agnes Beacon, and the valley of Trevaunance Cove with

Figure 7.3   Wheal Coates, St Agnes
Source: Photo by Hilary Orange

its beach and former harbour area. During the seventeenth to early nineteenth centuries, the St Agnes mines were some of the most important in Cornwall, and consisted mostly of small scattered mines, some of which continued activity into the early twentieth century (The Cahill Partnership and Cornwall Archaeological Unit 2002: 8–9). Mining remains visually dominate the village and the coastal area, and the latter contains one of the best known Cornish mine sites, the nineteenth-century workings of Wheal Coates which is now managed by The National Trust (Figure 7.3). The settlement of St Agnes developed upwards and outwards around its mines providing views and vistas of engine houses and chimney stacks at the bottom of driveways and within modern bungalow estates (Acton 2005: 76). Since the 1930s the mainstay of economic activity has been tourism and the village's picturesque nature and its wide range of shops, pubs and galleries have made it a favoured residential area and a dormitory village for commuters to nearby towns (The Cahill Partnership and Cornwall Archaeological Unit 2002: 11).

Minions (Caradon Mining District)   The small village of Minions, located within the Caradon area of Bodmin Moor, was the setting of the nineteenth-century copper boom which led to mass immigration to the area (a situation often analogized with the American Wild West) and consequently the formation of many of the moorland

Figure 7.4   Minions
Source: Photo by Hilary Orange

settlements. Major industrial activity had ceased by the early twentieth century. However, small scale quarrying continues as does the right of commoners to graze sheep, horses and cattle on the moor. Activity in Minions focuses on a pub, post office, tearooms and a small heritage centre and the surrounding moorland, which contains an unusual concentration of prehistoric and industrial sites, and attracts many walkers (Gillard et al. 2004) (Figure 7.4). Management of the historic environment is complex given ownership (Duchy, Council, private), a number of statutory designations (Area of Outstanding Natural Beauty, conservation area, Area of Great Scientific Value, Regionally Important Geological Site) and the needs of graziers, wildlife, agriculture, local residents and visitors. Alongside the World Heritage Site, the Caradon Hill Area Heritage Project (CHAHP) was set up in 2006 with the aim of maintaining and conserving the natural and historic environment and enabling people to benefit from the shared landscape and heritage (CHAHP 2008).

Methodology

Data for this study was collected through a questionnaire survey of local residents. Alongside demographic questions regarding place of normal residence, gender, age, time lived in Cornwall and place of birth the following questions were asked and additional comment invited:

1. Do you have any connection to the mining industries?
2. What does the term a 'sense of place' mean to you, if anything?
3. Does [the case study area] have a 'sense of place'?
4. How would you describe this place?
5. Did you know that [the case study area] is a World Heritage Site?

An age parameter (of 18 years and over) was set and to qualify as 'local' respondent's normal place of residence was restricted to the electoral divisions which encompassed each site in this case the electoral wards of Morvah, Pendeen and St Just, St Cleer, St Ive, St Agnes, Hayle North and Hayle South. Questionnaires were distributed through:

1. A drop-box system at public places within the electoral ward/s including libraries, shops, community centres, pubs and cafes. This sample is therefore stratified in terms of individuals who frequent these places.
2. A random postal survey of 100 residences within each electoral ward.
3. Placing questionnaires and stamped-addressed envelopes under car windscreen-wipers in on-site car parks.

In total 284 questionnaires were returned which fulfilled the parameters of the survey.[1] Definitions of sense of place were first categorized according to key word and theme and categorical variables were analysed through SPSS (Statistical Package for the Social Sciences) which allowed for the creation of both descriptive statistics (count and frequency) and inferential statistics (cross tabulation and chi-square tests).

Results

Demographics

To briefly describe the demographics of the respondents: 55% were female and 45% male; their age range was between 18 and 89 years of age and just over a half were aged between 30–59 years of age; the majority had lived in Cornwall for over 20 years; approximately a third had been born in Cornwall; 81% knew that the case study area was a World Heritage Site and around a quarter felt that they had

---

1 The following numbers of questionnaires were returned per site: Botallack 86, Hayle 59, Minions 67 and St Agnes 72; 106 were the result of postal survey, 152 through public venue collection and 26 through car parks (average of 27% response rate for the postal survey).

a connection to the mining industries. Of those who stated a connection to mining industries, 58% had been born in Cornwall. Comments revealed the different ways in which respondents were connected to the mining industries. Some referred to their 'ancestors' or to specific relatives (husbands, great grandparents, grandparents, uncles, fathers and in-laws) who had been miners at the mines of Geevor, Gwennap, Botallack, Levant and South Crofty or had worked at the foundries of Harvey and Co. and Holmans. Others had relatives who had emigrated to North America, South Africa or Malaysia. Some respondents had worked at Geevor during the 'live' period, or were former coal-miners, or had worked abroad in the industry. In the post-industrial period a few had been involved in restoration projects, had undertaken archival research or were involved in mine exploration.

What 'Sense of Place' Means

Responses to the question 'what does the term "sense of place" mean, if anything?' were categorized under a series of headings, the main ones being:

Place   'Historic heritage – affect of humans on landscape'. (Minions)
           'A unique atmosphere or physical manifestation found only in a particular
           place' (St Just)
           'The industrial heritage has shaped the two main areas of Hayle. (Foundry &
           Copper-House) and this dominates the townscape even today'. (Hayle)

The concept central to this theme is inherent qualities of place, whether tangible or intangible, which are framed as being external to the respondent. This includes notions of the spirit of place (referenced by some respondents to Lawrence Durrell's 1969 publication), sometimes expressed as 'essence' or 'atmosphere' and descriptions of particular aspects of the environment which make the place unique. The character of distinctive elements is such that as one respondent noted 'you couldn't be anywhere else', an impression reinforced by representations of iconic features. Such iconic features become defining elements, not just of the immediate locale, but of Cornish heritage and culture. Therefore the Crowns' engine houses at Botallack, or iconic sites such as Wheal Coates at St Agnes *are* Cornwall.

Belonging   'To me it's my home town having lived here all my life'. (Hayle)
              'Where you feel you belong, a place you feel belongs to you'. (Hayle)
              'Sense of place to me means Cornwall: One and All'. (St Just)

Here the term 'a sense of place' is described in terms of personal biography and social identities; place as the setting of the life story of the respondent – it is their home, it is the people that they know, it is memories of childhood and a feeling of connectedness, most commonly expressed as 'belonging'. For some, 'sense of place' is expressed in terms of the everyday aspects of life in the place. For example, one respondent described living in Hayle as a 'comfortable way of life [with the] pleasant company of neighbours' whilst another from the same town commented that he liked 'going out in the boat fishing' but apart from that, the town had nothing else he wanted.

Belonging can become a reciprocal arrangement; if a person belongs to the land, then the land belongs to them. As one respondent noted: 'I am Cornish; this is my county (St Agnes).' In informing a sense of identity place also connected to concepts of Cornish ethnicity which at its strongest can denote a nationalist (and anti-English immigration) ideology – this is implied within a contestable, but linguistically interesting, statement of Celticity from a respondent from St Just: 'for the real Britons, Welsh and Cornish, Land is everything; the word *hiraeth* in both languages (longing, homesickness, nostalgia) refers to place'.

Cognition   'Somewhere I can "feel" the history by observation and by reading'. (St Just)
        'A feeling of "this is where we came from"'. (Minions)
        'Peace, tranquillity and wondering what it must have been like in the 1800's'. (St Agnes)

This category incorporates responses which define 'sense of place' in terms of emotional response, knowledge and understanding. It includes generic emotional response, such as the rather fuzzy 'way it makes you feel', the emotional 'value' of the locale in terms of feelings of peace and 'rightness' (the latter interpreted to mean authenticity) and its historical qualities (being able to sense the past). The emphasis on sensory perception therefore embraces immediate and reflective impressions of place, for example, impressions of those who are new residents. It also incorporates an emergent appreciation of the historic environment through development of knowledge and understanding of the place. As one respondent noted, a sense of place is also about 'taking time to learn what has happened there in the past' (St Agnes). Therefore the cognition category connects to a continuum through which residents, whether of six months or 25 years, become active participants in the process of 'knowing their place'.

Other and Nothing   The 'other' category includes rare or ambiguous responses, including: 'In a sense I am reminded of that T.V. series *The Prisoner* – that's how I feel here at times' (Hayle); and 'Going through graveyards counting the young dead. Who cares for the family who are left with nothing' (St Just). Responses under 'nothing' include: '?' (Minions); 'Does not mean much to me' (St Just); and 'Frankly, little or nothing: it smacks of jargon' (St Just). Table 7.1 shows the overall count per category:

Table 7.1   Count of definition categories

Category	Frequency	Valid %
Place	73	29.8
Belonging	76	31.0
Cognition	39	15.9
Nothing	25	10.2
Other	6	2.4
Total	245	100.0
Missing	38	
Total	283	

Relationships Between the Data

Statistical tests were conducted in order to assess the probability of significant relationships between definition and other categorical variables (for example, age range, gender, connection to mining industries) with the aim of inferring more general, yet tentative, conclusions concerning the wider population, in this case residents of the wider World Heritage Site. The data for the definition category 'other' was omitted from the analysis due to a small count and the consequent effect on the validity of the tests undertaken. Therefore the following analysis refers only to data from the definition categories place, cognition, belonging and nothing.

Two-sample chi-square tests were used in order to test the suggestion (or 'null hypothesis') that there is no relationship between the definition of the term sense of place and gender, age range, connection to mining, place of normal residence,

Table 7.2   Chi-square test results – definition and other categorical variables

Variable	Asymp.sig (p) (2-sided)
Gender	0.957
Age range	0.692
Time Lived in Cornwall	0.260
Connection to Mining	0.221
Born in Cornwall	0.004

Table 7.3   Chi-square test – definition and born in Cornwall

	Value	Df	Asymp.Sig. (2-sided)
Pearson Chi-Square	13.217	3	0.04
Likelihood Ratio	13.080	3	0.04
Linear-by-Linear Association	.975	1	.323
N. of Valid Cases	207		

place of birth and the time lived in Cornwall. The probability results are shown in Table 7.2.

These tests demonstrate that, on the basis of probability, the alternative hypothesis that there is a statistically significant relationship between the way that residents will define a sense of place and their place of birth is supported (Table 7.3). Equally, there is no relationship between definition and gender, age range, time lived in Cornwall and connection to mining. The difference between the observed and expected count (Table 7.4) indicates a pattern which suggests that residents who were born in Cornwall are more likely to define a sense of place in terms of belonging and those who are not born in Cornwall are more likely to define a sense of place in terms of cognition.

When asked whether they felt that their place had a sense of place, 87% responded 'yes', 11% 'no' and 2% responded that 'they didn't know'. Paradoxically 58% of those who responded 'no' still defined the term according to the categories

Table 7.4   Definition and born in Cornwall cross-tabulation

			Born in Cornwall		Total
			Yes	No	
Definition	Place	Count	16	44	60
		Expected count	19	41	
	Belonging	Count	32	36	68
		Expected count	21	47	
	Cognition	Count	12	43	55
		Expected count	17	38	
	Nothing	Count	4	20	24
		Expected count	7	17	
	Total	Count	64	143	207

of place, cognition and belonging described above. More understandably, the remaining 42% felt that the term meant nothing.

Their Sense of Place – Descriptions of Place

The descriptions of case study sites presented above, written mainly using archaeological sources, provide one way of characterizing place – through time-depth, key features and economic/cultural activity. To provide comparison, counterpoint and exemplary material to support the definition categories described above, a montage of additional comments from respondents regarding their locale and their 'sense of place' follows. This commentary appears to support Alfrey and Putnam's (1992: 40) assertion that popular perception tends towards mythologizing 'affective (heroic, demonic or romantic) aspects of industrial culture'. Indeed, on the one hand place is portrayed in terms of beauty and aesthetics; ingenuity and skill; the romance of the ruin and nostalgia, whilst conversely place is an industrial wasteland, a memorial to dereliction, the past filled with terror and future change desired. Much of the commentary lends itself to romance, heroism and the aspirational and following Johnson's (2007: 17) argument that 'romanticism forms the backdrop to a large part of archaeological thinking about landscape' it could also be argued that these, the public's, senses of place are also filtered through a romantic lens. To emphasize these trends, the following comments from respondents have been deliberately polarized to give different senses of place according to romantic/heroic and unromantic/demonic themes.

Botallack   Keywords: Beautiful. Dramatic. Heroic. Interesting. Peaceful.

> 'Botallack is a coastal area with a unique blend of landscape and mining history which defines the county and is culturally relevant to Cornwall. Perched high above the sea are engine houses and amazing chimneys and enormous areas of mining waste. Little grows here due to the poisonous heavy metals but nature is gradually healing the scars of industry and it's a shock for visitors to see such an industrial landscape in a rural area. It reveals the history of mining, not only of hard lives but it's also evidence of man's ingenuity and pioneering efforts to mine in an unforgiving and dangerous terrain. It's moving to think of so many men going underground and out under the sea and to think of the hard lives of the women and children working on the surface.'

Keywords: Bleak. Boring. Gaunt. Harsh.

> 'Botallack is rather run-down and shabby and it has a preponderance of unsightly industrial ruins of little evident value; it's not like "proper" old places like ancient villages and stone circles. It's unfortunately changed since the removal of its spoil heaps during the 1970s by Geevor mine to recover the tin and copper still retained in them and it's also been spoilt by the restoration work and cleaning

up by the National Trust and the endless out-of-proportion mine collars. It's an industrial wasteland and it needs to be put back in working order.'

Hayle   Keywords: Beautiful. Comfortable. Diverse. Friendly. Old fashioned. Quirky. Safe.

'Hayle is a town with everything – it is a picturesque beach resort built around an estuary/river mouth, with good shops and convenient access to other parts of Cornwall via the A30. It is said that Hayle is a hard place to leave – it has a very close community and is friendly and down to earth. In the nineteenth century it was an industrial centre of importance; with two main areas of interest at opposite ends of the town ... built around the activities of the Copper Company and Harvey's Foundry. People often stay in one area only and this makes Hayle different to other towns. Hayle might have an industrial past but with the possibility of redevelopment its future lies in tourism.'

Keywords: Decay. Neglect. Poor. Shabby.

'Hayle is poor. It has no pronounced industry; it is poor in appearance and poor in employment and has lost its importance and most of its historic buildings and features. Most people who live here do not work in the town – it is a dormitory town and it's a poor relation to the neighbouring towns of Penzance and St Ives. Its sense of place was lost in the 1970s with the influx of tourism and people from outside the county coming to live here. Its sense of place is being lost due to too much redevelopment. Hayle is sadly neglected and is in desperate need of regeneration. Whilst there has been some improvement in Foundry Square, the quaysides are still rather derelict and there are many empty buildings. It has no real focus, apart from the two ends of Copperhouse and Foundry; there is nothing in the middle and it needs pulling together.'

St Agnes   Keywords: Beautiful. Evocative. Fascinating. Historic. Impressive. Natural. Scenic. Undulating. Windswept.

'St Agnes was one of the most active mining areas in its time and is of great historical interest. It has a past and a story to tell; it has lovely walks and is relaxing in all the seasons and great for the outdoor life whether on land or sea. St Agnes feels old, the area inspired the Poldark novels [Graham 1945]. It's a striking and unique landscape because of the number of mine buildings within the village as well as impressive ruins still withstanding storm and gale on the cliff tops and others ivy-covered in hidden coombes. Two minutes walk out of the village and some areas feel like a Victorian time capsule. It is easy to imagine the lonely walks the miners had walking to and from their work, the tracks and paths of which can still be seen. But the evidence of mining has become softened over the years of disuse and is becoming more overgrown by nature. Its sense of place comes from the contrast between its natural beauty and industrial heritage and the interwoven structures of mining and residential areas. Some of the engine houses have a magical quality and one cannot but marvel at how some of the structures were constructed without access to modern cranes and other machinery. They have become lasting monuments to the industrial past and provide some of the most beautiful and iconic images of Cornwall.'

Keywords: Hardship. Horror. Unsightly.

'St Agnes' beauty is tinged with horror due to the hardship that Cornish people had to endure to make a living. The mining areas are derelict, unsightly and miserable, especially the spoil heaps. Some of the old buildings require renovating as they look ramshackle. Its sense of place has been despoiled by an influx of second home owners.'

Minions Keywords: Atmospheric. Beautiful. Enigmatic. Inspiring. Quiet. Spiritual. Stark. Unspoilt. Untamed.

'Minions, the highest village in Cornwall, is set within a beautiful, natural landscape which is steeped in history from prehistoric to modern times. Its sense of place is made up of a remarkable combination of natural scenery, mining, geology, heritage and ancient history exemplified by E.V. Thompson's novel Chase the Wind [1977]. Low grade agriculture has spared the utilitarian ravages of modern farming methods which means that more history remains written on its face and the conditions of the remains are good enough to illustrate mining conditions and activities in the nineteenth century. The waste heaps are each a distinctive shape and each chimney is like a megalith. They look like wild barrows and standing stones. You can sense that people worked and lived here. Some of the richest copper mines in the world were here in the nineteenth century and because of their impact on Cornwall, the UK, the industrial revolution and hence the world, the moor helps to define much of south east Cornwall. It has a quiet mysterious, stark, eerie beauty which is bleak in winter but in the summer full of tourists. Its open spaces create a sense of freedom and make it an excellent place to walk and a relaxing place to wind down; it's also a good place for climbing, biking, exploring and sailing. On a clear day the views, even as far as Dartmoor, are spectacular.'

Keywords: Neglect. Sad.

'The Minions has a sad collection of once proud buildings in neglect surrounded by beautiful moorland. It played a significant role in industrial development but at great human cost.'

Summary and Conclusions

Local residents' understanding of the term 'sense of place' is connected to four major themes: the intrinsic character and atmosphere of place, a sense of belonging, emotional response, and knowledge and understanding whilst, for some, the term is meaningless or jargon. Many of these themes have a connection to the past – to history, heritage and genealogy. Upon statistical analysis it was found that definitions of sense of place do not have a significant relationship to gender, age, the length of time a resident has lived in the county or any connection to the mining industries. However, residents who are born in Cornwall are more likely to define a sense of place in terms of belonging.

There is no single understanding of what sense of place means. From the fuzzy 'way that place makes you feel' to strong declarations of Cornish ethnicity, there are shades of meaning in between. There is a concern regarding the relationship between a sense of place and belonging, with its potential for containing nationalist and therefore, anti-other ideologies. Within comparative regional contexts the term should perhaps be used by those within archaeology and heritage management with care (and a degree of explanation) to avoid possible hijacking of message and meaning. Furthermore, whilst characterizing place according to its historical development and description of key features will pose no challenge to those within archaeology and heritage management, there are always likely to be challenges in achieving a comparable degree of resolution where residents' home, family and work lives are concerned. Beyond generic terms such as pride, attachment and identity details, such as the man who likes going out in his boat (but doesn't like much else about his town) or the woman who finds support in her neighbours, are perhaps best accessed through ethnographic methods.

It is clear that Cornish identity is strongly connected to the former mining industry; it also appears clear that the majority of residents (Cornish- and non-Cornish born) appear to have bought into the regeneration myth. Herein lies a interesting paradox – on the one hand there is an awareness of Cornish history and the functional 'ugliness' of the living industrial landscape and on the other a post-industrial desire to clean-up and romanticize that landscape. Following Alfrey and Putnam (1992: 40) local people do appear to 'mythologize' their place through heroic, romantic and demonic 'aspects of industrial culture'. Furthermore, residents' sense of place is notably romantic, and it could be argued, aspirational. It is the best aspects of place and the place people would like to occupy in the future. The way in which sense of place is seen as being 'positive' is revealed in the comments on the *loss* of sense of place due to changing demographics. If sense of place is primarily perceived and communicated through romantic notions, a comparison can be made to critiques of landscape, which have argued that the word and its associative concepts carries connotations of aesthetics and the picturesque (Cresswell 2004: 11; Olwig 1993: 318–19). It could be questioned whether the residents' responses are, in part, a reflection of recent media stories regarding the World Heritage Site, or an expression of an affirmative relationship to place. The questions remain – is there room within definitions of sense of place for the romantic and the dysfunctional, ordinary and ugly? Are there better ways and words which could be used within research and public engagement on the relationship and meanings which people have for place?

Acknowledgements

I would like to thank the following organizations for their kind sponsorship of this research: Cornwall Archaeological Society, The Hypatia Trust, Cornwall Heritage Trust, London Cornish Association and The Trevithick Society. I am very grateful to the following individuals for all of their advice and support: Dr Andrew Gardner, Dr Sue Hamilton, Dr John Schofield and Professor Clive Orton and thanks also to Sefryn Penrose for commenting on earlier drafts of this chapter.

References

Acton, B. 2005. In and Around St Agnes Harbour. In *Around Perranporth, St Agnes and Portreath,* in B. Acton (ed.), Landfall Walks Books, no 16. Truro, Cornwall: Landfall Publications, 76–100.

Alfrey, J. and Putnam, T. 1992. *The Industrial Heritage: Managing Resources and Uses.* London: Routledge.

Brayshay, M. 2006. Landscapes of Industry. In R. Kain (ed.), *England's Landscape: The South West.* London: Collins, 131–53.

Caradon Hill Area Heritage Project (CHAHP) 2008. *Landscape Strategy Report.* Liskeard, Cornwall: CHAHP.

Cornwall County Council 2009. *Cornish Mining World Heritage.* [Online]. Available at www.cornish-mining.org.uk [accessed: 22 December 2008].

Cresswell, T. 2004. *Place: A Short Introduction.* Oxford: Blackwell Publishing.

Deacon, B. 2007. *Cornwall: A Concise History.* Cardiff: University of Wales.

Durrell, L. 1969. *Spirit of Place: Letters and Essays on Travel.* London: Faber and Faber.

Gillard, B., Historic Environment Service and The Cahill Partnership. 2004. *Cornwall Industrial Settlements Initiative: Minions.* Truro, Cornwall: Cornwall County Council.

Graham, W. 1945. *Ross Poldark* (1st in a series of 12 books). London: Ward Lock.

Harvey's Foundry Trust 2008. *Harvey's Foundry Trust.* [Online]. Available at www.harveysfoundrytrust.co.uk [accessed: 12 October 2009].

Hayle Area Plan Partnership 2009. Hayle Area Partnership. [Online]. Available at www.hayleareaplan.org.uk [accessed: 12 October 2009].

ING 2009. *Real Estate Portal.* [Online]. Available at www.ingrealestate.com [accessed: 12 October 2009].

Johnson, M. 2007. *Ideas of Landscape*. Oxford: Blackwell.

Noall, C. 1984. *The Book of Hayle*. Abingdon, England: Barracuda Books Ltd.

Olwig, K.R. 1993. Sexual Cosmology: Nation and Landscape at the Conceptual Interstices of Nature and Culture, or, What Does Landscape Really Mean?. In B. Bender (ed.), *Landscape, Politics and Perspective*. Oxford: Berg, 307–43.

Palmer, M. 1993. Mining Landscapes and the Problems of Contaminated Land. In H. Swain (ed.), *Rescuing the Historic Landscape: Archaeology, the Green Movement and Conservation Strategies for the British Landscape*. Proceedings of a Conference held at the University of Leicester January 1993. Warwick, England: Rescue, 45–50.

Palmer, M. and Neaverson, P. 1998. *Industrial Archaeology: Principles and Practice*. London: Routledge.

Payton, P. 1997. Paralysis and Revival: The Reconstruction of Celtic-Catholic Cornwall 1890–1945. In E. Westland (ed.), *Cornwall: The Cultural Construction of Place*. Patten Press and University of Exeter, 25–39.

Payton, P. 2004. *Cornwall: A History*. Fowey, Cornwall: Cornwall Editions Ltd.

Rowe, J. 1953. *Cornwall in the Age of the Industrial Revolution*. Liverpool University Press.

Schwartz, S.P. 2008. *Voices of the Cornish Mining Landscape*. Truro, Cornwall: Cornwall County Council.

Sharpe, A. 1992. *St Just: An Archaeological Survey of a Mining District*. Truro, Cornwall: Cornwall County Council.

The Cahill Partnership and Cornwall Archaeological Unit. 2002. *Cornwall Industrial Settlements Initiative: St Agnes*. Truro, Cornwall: Cornwall County Council.

The National Trust. no date. *Botallack Mine: Self-Guided Trail*. [Online]. Available at www.landsendarea.com/trails/botallack. [accessed: 12 October 2009].

The Prisoner. 1967–1968. British television series broadcast by ITV.

Thompson, E.V. 1977. *Chase the Wind*. London: Time Warner Paperbacks.

Thorpe, S., Boden, D. and Gamble, B. 2005. *Cornwall and West Devon Mining Landscape: World Heritage Site Management Plan 2005–2010 Summary*. Truro, Cornwall: Cornwall County Council.

Chapter 8

Maastricht-Lanakerveld: The Place to Be?

Anne Brakman

The city of Maastricht, which is located in the southern most part of the Netherlands, sharing a border with Belgium, is one of the oldest cities in the Netherlands, boasting a long and eventful past. It is the capital of the province of Limburg, which is wedged in between Belgium and Germany. Human occupation of the area is known from ca. 350,000 years BC. Providing water, sufficient high ground and fertile soil Maastricht and its surrounding area has always drawn people towards it. During the Roman period, a bridge was constructed across the River Meuse to allow troops to move swiftly from northern Gallia Belgica to Cologne and the *limes*. This bridge (and its replacements) remained the most northern crossing over the Meuse for centuries. Due to this main road and the bridge, Maastricht evolved from a small *vicus* and *castellum* into a religious centre, concentrated around the basilica of Our Lady and the St Servaas basilica. Because of the strategic location of the city, Maastricht was often besieged and controlled by foreign occupiers, like the Spanish and the French, for long stretches of time. In the early nineteenth century Maastricht rapidly became a city of industry, well-known for its production of ceramic wares, from where the industrial revolution spread throughout the rest of the Netherlands. Today the city is an administrative centre and part of the Euroregion and is mostly known for its cultural and historic richness, which makes inhabitants and visitors alike feel at home there.

Within the boundaries of the municipality of Maastricht my colleague Gilbert Soeters and I are responsible for all matters concerning archaeology. Adjoining the border with Belgium, in the northwestern part of the municipality, is the Lanakerveld. At the moment, this is an agricultural area with a few farms, some dirt roads and two small streams but there are plans to develop it into an industrial zone and a new neighbourhood, separated by a green strip. The industrial zone will expand into Belgium, giving it an international aspect.

Prior to development, archaeological investigations took place on both sides of the border. Through desk-based assessment, a combined environmental and archaeological borehole survey, fieldwalking and trial trenches 19 archaeological sites have turned up on the Dutch side of the Lanakerveld alone. It is also clear that some of these find-spots continue on to the other side of the border between Belgium and the Netherlands.

Making sure that these places are properly excavated and documented before development starts will be a costly undertaking. Further, we were challenged by the client to try to find a way in which archaeology could enhance the quality of life

**Figure 8.1   The Lanakerveld as it is today: farmland with a small chapel;
the only readily visible cultural heritage in the area**
Source: Photograph courtesy of Mr S. Minis

in the Lanakerveld and lure potential homebuyers. We accepted the challenge, but
when it sank in, we realized it was not going to be as easy as putting up a showcase
with some potsherds in the community centre. We started asking ourselves: how
can archaeology and history make a place more appealing to future inhabitants?

Sense of Place

Before answering that question, I will try to define what is meant by '*genius loci*'
in order to better understand what it is we are after. The 'spirit of a place' is a
term which can be traced back to Roman culture where it was utilized to name
the atmosphere and the uniqueness of a certain location. This spirit is reflected
in 'the story of a place', which underlines that immobile heritage is the result of
development and interventions that took place in the past. Although history seems
to impinge a sense of connection on inhabitants of a historically rich location, it is
important to realize that *genius loci* is more than just history. Other defining factors
are the present era and the background of the inhabitants, to name but two.

Perhaps *genius loci* can be best described as the typical nature or atmosphere of a place that leaves an impression. The '*je ne sais quoi*' that distinguishes one place from the next and the emotional attachment people have to places they hold dear. Atmosphere, '*je ne sais quoi*', emotion, people ... how vague can the subjects of a sentence be?

Rather than attempt a definition here, a series of questions highlights some of the issues and the difficulties with creating a single definition of *genius loci* or sense of place. For example: How big or small does a place have to be to create a sense of place? Is it the entire city? Is it a square, a street, a church, a house? Is the sense of a place subject to the weather conditions or the time of day? Or is it a continuous presence, unaffected by space and time? Then how can it be that, on turning a corner, the sense of place we just experienced is gone? Is there some form of interaction between the onlooker and the place? Do only places with historic remains generate a sense of place? Does there have to be a certain time depth or can hypermodern objects and structures generate a sense of place instantly? Is it enough to make history noticeable and recognizable? Can it be created at all? Can we pass on this sense to others and if so, by what means? Is it all just a matter of perception? How do people pick up on a sense of place? Does it come naturally to everybody or is it something you have to be sensitive to? Is it more nurture than nature? Can people be immune to it or does it come naturally with places that are part of our personal history? Was there a stronger sense of place in the past than there is now? Or is a sense of place something that belongs in the present, created by a subconscious feeling that everything is changing too fast? Is that why we cling to the few remains we have left? And does this mean everybody can have an individual perception of sense of place? We create more and more environments that are focused on comfort and ease, but where we do not feel at home. Is it necessary for us to feel a sense of place in order to be happy or can we do without it?

It is hard to find a single meaningful and useful definition of sense of place. *Genius loci* is no more than a metaphor, not really something that could help create a solid social basis for archaeology or something spatial planners can apply in the design of a new neighbourhood. Surely, most will indicate they know what *genius loci* means, yet it is hard to define precisely.

Creating a Sense of Place

Perhaps it helps if we try to recognize when a sense of place, the feeling of *genius loci*, emerges and try to analyse and explain why a place has this sense. There are places where we are fed information or which have articulation like a *lieu de memoire* or site of memory. At these places *genius loci* can occur instantly. Here one expects to instill a certain feeling, perhaps based on a historical understanding, or an emotional attachment to the place. I will try to explore some places in my own town, Maastricht, on this basis recognizing the significance of historical depth in generating a sense of place amongst that place's inhabitants. I will start with some attempts that have been made to visualize the historic significance of a location, but turned out to be unsuccessful. First, there is the column known as the Obelisk. It consists of a pillar with a lion on top, a copy of one of the statues that was found among the remains of the Roman bridge over the River Meuse. This was erected beside the Stichting Romeinse Brug (Foundation Roman bridge) and represents the location where, in the Roman period, the bridge over the Meuse reached the western bank. It offers no connection to this location however, since, despite the grand presentation, hardly any inhabitants know of its existence, let alone tourists. This is mainly caused by the lack of people passing by. At this moment the location is separated from the busy city centre by a two-way street without level-crossing. Perhaps the promenade along the Meuse bank that is currently under construction will change this.

Then there is the art piece on one of the locations where a Roman road used to run through the city. This piece of art is part of a wider series that are on display in several cities within the northern part of the former Roman Empire. The art piece is a positive and negative relief of human figures. Hardly anyone makes the connection to the Roman road that used to pass this point. To the investigative observer there is a small plaque to be found on the back of the relief. Yet, this does not provide much useful information. It only mentions a few cities, and does not explain what is so special about these cities or what they have got to do with this relief. There is nothing at all about the purpose of the piece itself. Another visualization of Roman remains is the markings in the pavement outlining the Roman *castellum* wall in the Houtmaas. Again, there is no information explaining what this is and since it is in a parking zone, it is often obscured by parked cars.

The fourth example is the remains of the first medieval city enclosure (construction finished in 1229 AD) in the garden of the City Archive. Although publicly accessible, hardly anyone knows it is there and since it is not visible from the street, hardly anyone ever will. Then there is the tenth arch of the medieval Servaasbridge (Figure 8.2). During recent renovation of the Servaasbridge remains of the original thirteenth-century bridge were found. These remains include part of an archway, poorly preserved since they were partly demolished when the canal between Maastricht and Liège was dug. Nonetheless, inhabitants insisted they should be preserved and integrated in the new promenade. The result is disappointing. You have to be lucky to even get to see the remains, since the glass shield covering them is dirty and at such an angle that it reflects sunlight. There is no information whatsoever to explain what it is.

During the excavations preceding the construction of the new library Centre Céramique the foundations of the Parma-bastion were uncovered. Although there was a strong lobby to preserve these remains *in situ* and integrate them into the library somehow, the construction plans could not be altered to make this possible. In order to visualize this bastion in another way and remind people of the defences that were erected around the city in the eighteenth and nineteenth century, before they were torn down to allow the city to expand, it was decided to create an outline

Figure 8.2   The tenth arch of the St Servaasbridge, reconstructed in aluminium plating

of the ground-plan of the bastion in the asphalt and pavement. This seemed like such a good idea, but there were problems. For instance: there is no overview from ground level, so people hardly recognize it as an outline of something. There is an information shield to indicate it, but it is hard to find. At first the outline was often mistaken for a level-crossing and caused serious danger to traffic, which led municipal officials to darken the colour of the outline. This might be somewhat safer, but it has not helped increase the visibility of the bastion.

Finally, there is the De Spaansche Tombe (Figure 8.3). This is a preserved medieval motte. Mottes are artificial hills with a tower or castle on top for defence and to gain strategic advantage. They were erected from the tenth century until the later part of the thirteenth century, when they were replaced with other forms of defence. Although most passers-by notice it and there is some information on a shield, it is hard to get a realistic view of the motte as a whole since it is largely overgrown with shrubbery and trees. Recently, most of the shrubbery has been removed to improve the visibility of the motte, yet the trees still remain.

There are a number of other examples in Maastricht though of much more effective attempts at visualizing the past. One of these is in the basement of the Derlon Hotel. Here archaeological remains of a Roman sanctuary and the late third century Roman *castellum* are on display *in situ* and you can dine amidst the Roman gods! On Sundays it is open to the general public. Op De Thermen is a small square where differences in the colouring of the pavement indicate where

**Figure 8.3   The Spaansche Motte as it is today. It is just a matter of time
                  until nature renews her claim on it**
Source: Photograph courtesy of Mr S. Minis

the Roman bathhouse was. This seems to gain more interest from people than
the Parma-bastion. Perhaps it is because there is an overview of the bathhouse
from ground level. There is also the name of the square, which implies there is
something historic about this location and it invites people to investigate. They
might have some trouble finding the information shield, but at least there is one.

Then there are the nineteenth-century reconstructions of the medieval city
enclosures like the Waterpoortje, the Helpoort and Recentoren. It does not seem to
matter to people that these reconstructions are not very realistic. They are accepted
as old and genuine. The same goes for the Servaasbridge, which is a reconstruction
of the thirteenth-century bridge, but is none-the-less regarded as an authentic
feature of the city. Finally, there is the Landmarks project, an international project
between the Netherlands and Belgium which links certain points in the landscape
by trekking or cycling routes. Every point is easily recognizable because it is
defined by a huge rock. At every location there is information about a specific
event or period of time and its relation to the landscape; it is a good combination
of information and entertainment, or 'infotainment'.

Is it possible to distil from these examples of visualization approaches that
work well and approaches that do not? It is not surprising that when people do not

Figure 8.4   The Father Vink tower, part of the first medieval city enclosure. The romantic reconstruction shows battlements; these were not part of the original city wall, but conform to the nineteenth-century image of medieval defences
Source: Photograph courtesy of Mr S. Minis

know about something, it does not work so well. If it is not on a regularly used route, it will soon be forgotten and in some cases even vandalized. What seems to contribute to the success of some of these examples is that they are part of the whole; they fit into a larger context. People consciously look for expressions of this and expect a confrontation with the past. Even when you are not consciously looking for it, there is often an automatic reaction. An atmosphere is created and you become aware that you are a participant in something bigger.

Except for the Landmarks, all successful examples are located within the historic centre of the city. So what makes the Landmarks project work? They form a chain through connected landscapes with a mutual history. All facets of the history of the landscape are illustrated as layers of a palimpsest, from the geological origin to the sub-recent (from the sixteenth century up to the nineteenth century) sieges of Maastricht. The brilliance lies in the fact that the separate Landmarks give information on the location at hand and yet they invite visitors to explore the other Landmarks. The opportunity for this is offered by the cycling and trekking routes that connect them.

It is remarkable that authenticity is not a necessary factor for success, nor will success be imminent if archaeological remains are simply put on display to tell the story. It seems sufficient to create the idea of authenticity or to visualize the past in such a way that people are tantalized enough to want to explore for themselves the facts of what took place here.

Sense of Place in the Lanakerveld

So, returning to the Lanakerveld, this is part of a historically rich landscape, but how can we represent its history in a way that includes present and future inhabitants without the eye-catching medieval churches or castles that exist in central Maastricht? The time depth that the historical city centre naturally offers is missing here. A forced and fast transition in the use of a location, as for example the development of an agricultural area into a new neighbourhood, asks a lot of people's power to adapt. A way to smooth the transition is to show that the past of this landscape is not erased by the new layout and actually this new function is no more than the newest chapter in the history of that location. This is where the perspective of archaeology can be particularly helpful.

As mentioned earlier, the Lanakerveld is an archaeologically rich area, with at least 19 sites dating between the Neolithic period and the early Middle Ages. But with such richness comes the risk of information overload. Knowledge of what was going on in the Lanakerveld a couple of thousand years ago might gain appreciation from some but it is not enough to create a special bond with the landscape. Reconstructing a Neolithic farmhouse might help people to visualize the past, but it is unlikely to enable people to form their own image of the past, one that appeals to their personal sense of place.

So in an attempt to present the rich history of the Lanakerveld in a way that makes it visible to the inhabitants and linked to their personal history, and rather than force-feeding them all the information at hand, we proposed an alternative scheme. We were inspired by the footprint path that has been used by Gerrets (2008) to accentuate the findspot of a Neolithic grave in Hardinxveld-Giessendam (see Figure 8.5); we took this idea a step further. Instead of using only the imprints of bare feet, we decided to vary the imprints according to time and use them in a way that allows people to step almost literally in the footprints of their ancestors. Bare feet would be used to convey the times when man first came to the Lanakerveld, leather covered feet for the Bronze and Iron Age, from circa 3000 BC to Roman times, Roman sandal imprints, medieval shoes, soldier's boots for the subrecent sieges and the Second World War and clogs for today's farmers. The path can be stretched throughout the entire Lanakerveld, maybe passing some specific accentuated points at the locations of sites. The imprints from the different areas could also be linked to the sites; perhaps changing not only through time but also by site. It should however lead to a public place in the centre of the community and end with the imprints of the new inhabitants, thus creating a link to their own history and visualizing a lasting memory for the new inhabitants of their role as pioneers in this new settlement.

**Figure 8.5   The bare foot imprints in Hardinxveld-Giessendam that lead
                up to the location where the grave of 'Trijntje', a Neolithic
                woman, was found**

Source: Het verhaal verbeeld, Paul de Kort, Parklaan landschapsarchitecten and RAAP
2008: 30.

Due to the global economic downturn, it was decided in 2009 to postpone the
development of the neighbourhood in the Lanakerveld, which could mean a delay
of 20 years for the realization of the neighbourhood or even a cancellation of the
planned development altogether. This means the archaeological remains can be
preserved *in situ* for the time being and ideas concerning the visualization of the
rich past of the Lanakerveld can be specified in more detail. Currently, the concept
of attempting to create a *genius loci* by using archaeology and history is being
integrated into our visions for existing cultural heritage, and for archaeological
remains still concealed in the ground. It is no longer the main goal of cultural
heritage managers to simply visualize archaeological remains by brushing them
off, or using building reconstructions; now they consciously ask themselves how
to visualize it in a way that appeals to the onlooker.

Conclusion

Although it is hard to find a comprehensive definition of '*genius loci*', it has become clear that history and archaeology contribute to inhabitant's sense of place and their wellbeing. Increasingly this is becoming a topic of interest, not just affecting archaeologists and historians, but project developers as well who are realizing that the contribution of archaeology and history to quality of life is considerable. When people feel connected to a certain place, they are more likely to live, work, visit and shop there.

Although such a connection is often created naturally when people spend time in a place, linking their personal history to that of the place, it seems hard to create this connection instantly. But instead of expecting an instant sense of place by brushing off archaeological remains and putting them on display or providing an overdose of information about the past, it is important to try and find a way in which the history of the location links to the inhabitants' personal world of perception. By providing the future inhabitants of the Lanakerveld with a literal guideline they can create their own image of the past and feel a part of it.

Acknowledgement

I would like to express my gratitude to Gilbert Soeters for his contribution to this chapter.

References

Gerrets, A. 2008. *Het verhaal verbeeld, Zes schetsontwerpen voor archeologische vindplaatsen in Zuid-Holland*. Hardinxveld-Giessendam: Provincie Zuid-Holland.
Groffen, B. and Posthuma, A. 2008. *Levend verleden, Een reflectie op inzet en handelen cultuurhistorische sector in Leidsche Rijn e.a. casussen*. Utrecht: Projectbureau Belvedere.

Chapter 9

Between Indigenous and Roman Worlds: Sense of Place in the North-Eastern Iberian Peninsula During the Roman Period

Paula Uribe

The lure of the local is not merely a recent invention. It is evident also in Roman culture and most probably originates much earlier than this. Roman politics of conquest ensured that their sense of place was carried throughout the empire. The organization of the city worked as a model of control, which transported the Roman way of life to all the new territories, including the Iberian Peninsula at the beginning of the third century BC. Thus, we can see how the Roman conquerors reproduced a 'little Rome', the metropolis, in every *ex novo* foundation in the new territories.

A civilization's sense of place is often best reflected in domestic architecture. This chapter examines a small group of examples belonging to the most important place for a person and his family during the Roman period: the house. Such is the importance of the house as a physical support of the Roman family, that *Publius Syrus* affirms in his *Sententiae* (182) that: '*The exile with no home is a dead man without a tomb*'. During the Roman period, houses expressed sense of place in two different ways, linked to the significance of public and private life in Roman culture.

In the private sphere, from *The Odyssey* onward, throughout life and literature, the *oikos* of the Greeks and the *domus* of the Romans provided a strong focus of emotion (Treggiari 1999: 33). The owner and other residents might acknowledge emotional ties to residences. Cicero explicitly states his affection for the family's old home near *Arpinium* and the countryside in his *Letters to Atticus*:

> Why should I invite you to Arpinium? A rugged place, but a good nurse of boys, nor can I look on anything sweeter than that land. (Ad Atticum 32/II.11.2)

Familiarity is again the reason for Cicero's affection in the passage on *Arpinium* in his *On the Laws* (Treggiari 1999: 36):

> I do come to this beautiful and healthy place when I can get away for several days, especially at this time of year, but I hardly ever get the chance. I have another reason for liking it, which doesn't apply to you, Titus. What is that? Because it's

our own country to my brother here and me. More than that – you see this villa, now rebuilt in more luxurious style by our father, who lived most of his life here in literary pursuits because of his weak health. While my grandfather was still alive and this was a small, old fashioned villa, like that of Curius in Sabine territory, I was born in this very place. So there is something deeply hidden in my mind and feeling, which perhaps makes me like this place more, just as that wise man, as the poem says, gave up the chance of being immortal just so he might see Ithaca again. (Legg 2.3)

Cicero's house and birthplace were places full of meaning and emotion, which he compared to the mythical Odysseus' home, who renounced immortality to return to his *oikos*, in Ithaca. A sense of home as a refuge from troubles is strongly reflected by Cicero when he ponders country retreats.[1]

Familiarity is not the only emotion linked to dwelling and a sense of place. Martial, a Hispanic epigramist,[2] reveals to us his conception of an ideal life by describing the characteristics of the house needed to achieve this ideal:

> II, 90.
> Quintilian,[3] supreme guide of wayward youth,
> Quintilian, glory of the Roman gown, forgive me that I a
> poor man and not crippled with years, am in haste
> to live. No man is enough in haste to live. Let him
> put it off who prays to surpass his father's riches
> and crowds his hall with excess of portraits. My
> pleasure is a hearth, and a roof that does not resent
> black smoke, and a running stream, and fresh grass.
> Let me have a well-fed, home-bred slave, a wife not
> over-educated, the night to sleep, the day without a quarrel.

But there is a public life that every Roman aristocrat had to achieve in order to succeed in a political career. This public sphere was also reflected spatially in the places they chose to live. Cicero first lived in a family house on the *Carinae*, and later moved to the Palatine, an area full of senatorial houses. He transferred to a place where his house would be better regarded, in accordance with his new status and more appropriate for his equestrian father and for himself, a candidate for the highest office (Treggiari 1999: 36 n. 18).

---

1 On the concept of the countryside as a refuge from troubles see: Martial XII, 57.

2 *Marcus Valerius Martialis* was a Latin poet writing in the first century AD. His place of birth was *Augusta Bilbilis* (now Calatayud) in *Hispania Tarraconensis*. He is best known for his 12 books of *Epigrams*, published in Rome between 86 and 103 AD, during the reigns of the emperors Domitian, Nerva and Trajan.

3 *Marcus Fabius Quintilianus* (c.35–c.100 AD) was a Roman rhetorician from Hispania.

Sense of place transcended the locality chosen to live in; inside the house, the concept of place is expressed through the use of different materials. As *Petronius* wrote in his *Satyricon*:

> The visitor to Trimalchio's house was confronted by a succession of signs, a mute but eloquent code which pointed past the fabric and subordinate personnel of the domus to the dominus himself, creating an impression which not only reflected on the standing of Trimalchio but conspired to enhance it. The green and red-clad porter, shelling peas in the entrance into silver bowl; the golden birdcage suspended above the threshold; the starling watch-dog painted by the porter's cell, followed by biographical frieze representing the master's rise to fortune; the shrine displaying silver Lares, a marble Venus and golden box: all these were a prelude to the approach to the triclinium where ultimately, after much further ado, the great man would greet them. (Wallace-Hadrill 1988: 44)

Cicero concludes (*De Officiis* i. 138–9) that a house can be an ornament to dignity, and *dignitas* needs *aedificatio* to aid the *dominus* in bringing his *domus* to stand as much as the other way around. In this way, the *dominus* defined his own sense of place inside his house, transmitted through different signs belonging to a local language.

This is the way in which a sense of place is reflected in Roman houses. In this case, the consumer, through his expenditures, transmits signals in a language of social communication. Like any other language, as Wallace-Hadrill (1988: 48) states, it has its own grammar, rules and signs. Reading this language in the archaeological evidence, we can gauge how private architecture during the Roman period was one of the vehicles of social promotion and, at the same time, the reflection of the materialization of the *dominus* himself, reflecting both propaganda and emotions.

Below are different cases from the northeastern part of the Iberian Peninsula, which was conquered by the Romans between the third and first centuries BC. The conquest resulted in an early settlement of Italian colonists, who transferred their *modus vivendi*, and with it their sense of place, into the Hispanic lands.

Colonia Iulia-Lepida Celsa (Velilla del Ebro, Saragossa, Spain)

The city of *Celsa* was the first colonial settlement founded by Romans, with a privileged legal status in the Ebro valley. It took its name from the *cognomen* of its founder *Marcus Aemilius Lepidus*. Very likely it was founded c. AD 44, during his second mandate in Hispania, while being proconsul of *Hispania Citerior*. It seems that in c. AD 36, the triumvir Lepidus' fall from grace motivated the suppression of his name in benefit of the old name of an Iberian predecessor *Kelse*, and was called from then on *Colonia Victrix Iulia Celsa*.

The creation of this new city of privileged status revitalized this geographic area, the central sector of the Ebro River, where the majority of the urban settlements

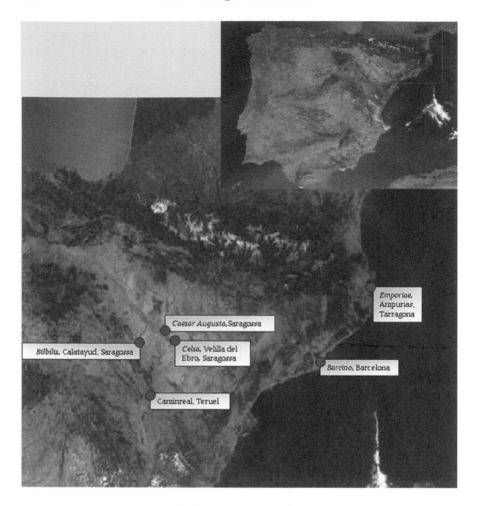

Map 9.1   Map showing several Roman colonies in the North-East of the Iberian Peninsula

had been systematically devastated and depopulated during the Sertorian War. The inhabitants of the colony were Italian immigrants and veteran legionaries, who as citizens received a proportional part of the land.

Among the excavated Roman houses there are two buildings in particular that stand out: the House of the Dolphins and the House of Hercules. The House of the Dolphins was the result of the union of two houses in the late Augustan period (1 and 14 AD), characterized by typical Italian decorative features and design. The pavements, the decoration of the walls, the kitchen and a small *cubiculum* or bedroom, are also similar to the Roman houses of Pompeii and Herculaneum.

With regard to the House of Hercules, special mention should be made of two rooms that showed the visitor the power of the family that lived there. Specifically, room number five is famous for its paintings (Second style[4]) where the mythological story of the labours of Hercules were recreated, reflecting the influence of *luxuria asiatica*. A space (14) was opened and decorated with two columns adorned with Corinthian capitals in the central axis of the peristyle. This room was also flanked by two rooms of similar depth, which are very correctly related to the tripartite division of the space. The atrium and the peristyle connected them with pure Roman tradition.

Emphasis is laid on these living rooms. They are preceded by two columns, and their great sumptuousness reminds the viewer of the Roman public basilica, conceived for the contemplation of the gardened peristyle. The use of public architecture inside the private home is a clear symbol of power.

Emporiae (Ampurias, Catalonia, Spain)

The first example of typical Roman colonial architecture is the outstanding village on the eastern coast of Hispania, a place originally known as *Emporian and* founded by the Phocaean Greeks c. 600. The city extended over the little island of *Palaiapolis* in the Gulf of Rosas. Enriched by the commerce between the Greek colonies of the South of France and the Tartessian centres of southern Spain, the enclave very soon became a flourishing inland colony.

The alliance of this colony with Rome against Carthage converted *Emporion* into a strategic centre with a permanent *praesidium* located some metres from the Neapolis. From then on, Ampurias became the epicentre of Roman influence in the Iberian Peninsula.

Houses 1, 2A and 2B (Figure 9.1) are examples of buildings owned by high status Roman individuals; according to the rules recommended by Vitruvius,[5] (VI, 5, 1): '*for those who have to give audience to the citizens, magnificent halls have to be built, high atria, spacious peristyla, gardens and avenues ...*' These are closely similar to the houses defined by Gros (2001: 72–3) as characteristic of the Republican senatorial oligarchy. As in the previous examples, these buildings are characterized by their extended area and large dimensions, especially in comparison to other nearby houses in the Neapolis. Their extensions included vast landscaped areas, related to spacious summer *triclinia* (or dining room). Outdoor dining rooms have been documented, where pleasure and ostentation accompanied the celebration of *cenae* (or dinner). In the same way, these spaces were decorated

---

4   The Second style, or architectural style of Roman mural painting, dominated the first century BC.

5   Vitruvius (born c. 80–70 BC, died after c. 15 BC) is the author of *De architectura*, known today as *The Ten Books on Architecture*, a treatise written of Latin and Greek on architecture, dedicated to the emperor Augustus.

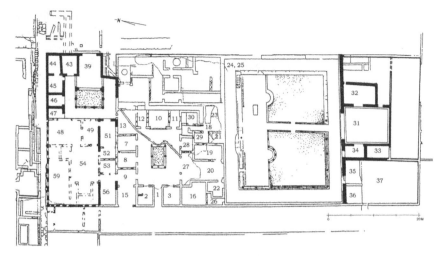

Casa 1 o Casa de Villanueva

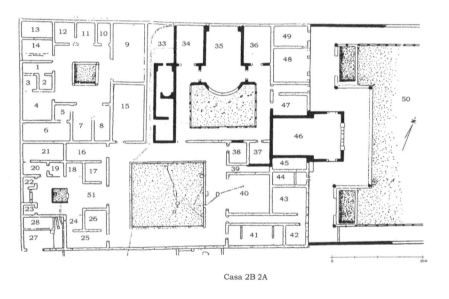

Casa 2B 2A

**Figure 9.1   Plan of houses 1, 2A and 2B from *Emporiae*, Ampurias,
          Tarragona, Spain**

distinctively: with mural paintings, *opus signinum*, white and black *tesellatum* and
emblemata vermiculata, similar to the Faun's house in Pompeii.

This kind of residence, with larger dimensions and rich decoration, very quickly
spread all over Italy, mirroring those of the senatorial Roman oligarchy as closely
as possible, like Cicero's house on the Palatine. The earliest case of imitation was

documented in *Fregellae,* a colony founded in 328 BC, where houses with big Tuscan *atria* were built.

The houses in Fregellae come closest to Gros' definition (2001: 72–3) of Republican senatorial oligarchic houses. The similarity between a public and a private building enabled high status figures to exhibit their great power.

It follows that these houses were occupied by Italian settlers, who belonged to high social classes and were involved with maritime commerce. In addition to this, the works performed in the residences – their configuration as a single house or the introduction of a *balneum* (or bathroom) – indicate the growth of their wealth during the years.

La Caridad (Caminreal, Teruel, Spain)

The next example is a city of unknown name, founded *ex novo* and built following a regular plan. The reliability of its archaeological documentation and the morphology of the buildings make it the best example to illustrate the redefinition of place after local indigenous elites tried to adopt the Roman style of life.

Figure 9.2   Drawing of a pavement at Caminreal

We can infer from the current data that there was a single period of very short occupation of Caminreal between the second century BC and the first third of the first century BC. The so-called Casa de *Likine* of Caminreal (Vicente et al. 1989: 30) dates to this time and was organized around a central court with portico, without any documented *impluvium*, or cistern, and a central luxurious room. It is a simplified model of the peristyle house, which was later developed in the Iberian Peninsula. Inside the building, complete Italian decorative repertories may be found, occupying the central room, famous for an inscription in the Iberian alphabet.[6] The transcription revealed the name of the owner – *Likine*.

It is interesting to note that a member of the local elite had chosen to represent himself in this way, adopting very quickly all the Roman symbolic concepts of power in architecture. Elite houses were conceived of as completely Italian; however, they did not abandon their own language – Iberian.

Municipium Augustam Bilbilis (Calatayud, Saragossa, Spain)

A similar example is found in *Bilbilis*, which shows, even more clearly than the previous examples, the adoption of different characteristics of Roman domestic architecture, used as propaganda by the owner. Situated on the shore of the Jalón River, the city of *Bilbilis* originated from a transduction (the establishment of Italian colonists in an indigenous village). Its indigenous name was changed to *Bilbilis Italica*, from the beginning of the first century BC onwards, with Italian settlement from those dates. From the Augustan period, a typical Roman city was designed, with a big monumental centre which comprised a forum, temple, basilica, curia, porch and cryptoporch; in addition to *thermae* and theatre (Martín-Bueno 1975, 1987, 2000).

Among all the houses, *Domus* 1 deserves a special mention. It was part of a group of houses which together comprised the *Insula* I of the so-called *thermae* district, located in the north-east of the city, in the same natural terrace as the public *thermae*. Among the four *domus* of the *insula*, the *Domus of the Balneum* preserves the most characteristic Italian ground floor, because it was planned as a house that organized its rooms and spaces around a central tetrastyle *atrium* (an atrium with four columns).

The house was internally divided into three floors arranged around the tetrastyle atrium. The representation spaces or public places were placed on the west side of the second storey, interpreted as the noble floor, which was divided into three rooms: the central one occupied by the *tablinium* (the office of the *dominus*), decorated with a pavement of *opus signinum* in rose shape; the *triclinium* (or dining room) in the south-eastern side with pavement of white terrazzo; and the *balneum*

---

6   Transcribed in the following way: *l.i.ki.n.e.te. e.ki.a.r. u. se.ke.r.te.ku.* (Vicente et al. 1993: 755).

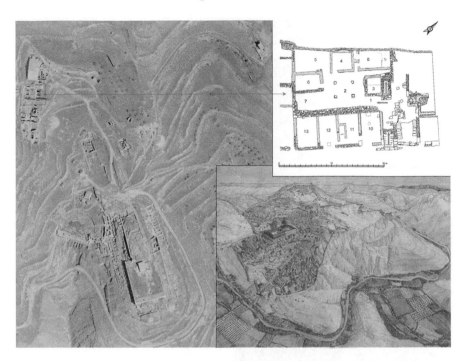

Figure 9.3   The town of *Bilbilis* (Calatayud, Saragossa, Spain) and plan of *Domus* 1

in the north-eastern corner, even though it was built during the refurbishments performed in a second stage of the *domus*.

The presence of a thermal installation in the house reveals that it belonged to a member of a very rich social sector, because its construction and maintenance demanded great expense and the public *thermae* and *foricae*, or public toilets, were only located on the other side of the street. But, in this case, it represented a perfect example of the wish of the Italian immigrants to transfer a typical Italian structure to the Iberian Peninsula, where indigenous houses did not have bathrooms.

In contrast though, there is the example, of an indigenous inhabitant of Bilbilis, the famous Hispanic poet Martial, who felt homesick for his home in Bilbilis when he had to move to the capital of the empire, Rome. When he moved to the capital for the first time, he considered his home as *provincialis solitude*, but 34 years later, when he returned, he wrote the following words:

Figure 9.4   The *Balneum* of *Domus* 1 from *Bilbilis*

XII, 31

This grove, these founts, this matted shade of
arching vine, this conduit of refreshing water, and the
meadows, and the beds of roses that will not yield to
twice-bearing Paestum, and pot-herb in January
green, nor seared by frost; and the tame eel that swims
in its shut tank, and the white dove-cote that harbours
birds as white-these are my lady's gifts: to me
returned after seven lustres has Marcella given this
house and tiny realm. If Nausicaa were to yield me her
sire's gardens, I could say to Alcinous 'I prefer my own.'

XII, 18

While perchance you are restlessly wandering,
Juvenal, in the noisy Subura, or treading the hill of
Queen Diana; while, amid the thresholds of great men,
Your sweaty toga fans you, and, as you stray, the
Greater Caelian and the less wearies you, me my

Bilbilis, sought once more after many Decembers, has
received and made a countryman, Bilbilis, proud of its
gold and iron. Here indolently, with pleasant toil, I
frequent Boterdus and Platea (such in Celtiberian
lands are the uncouth name!); I enjoy a huge uncon-
scionable sleep which often not even the third hour
breaks, and I pay myself now in full for all my sleep-
lesness for thrice ten years. Unknown is the toga;
rather, when I ask for it, the first covering at hand is
given to me from a broken chair. When I get up, a fire,
served with a lordly heap of logs from neighbouring
oak-wood, welcomes me, and my bailiff's wife crowns
it with many a pot. Next comes my huntsman, and he
too a youth whom you would desire to consort with in
some secret grove. The unbearded bailiff gives my
slaves their rations, and asks permission to crop his
long hair. So I love to live, so I love to die.

Martial exaggerates and idealizes his home after returning from Rome. His
emotional attachment to his birth-place changed when he started feeling nostalgic,
as in Cicero's case. Before his visit to Rome, Martial felt that no one understood
his poems in his provincial town of Bilbilis. When he returned from Rome, he
realized that his native town was synonymous to a quiet life, abundance of food,
and deep sleep.

Conclusions

People acquire different semantic categories to give a place meaning if feeling
rootless and homesick. In the case of the Roman houses in the Iberian Peninsula,
their inhabitants fought the feeling of nostalgia by transmitting some physical
characteristics and semantic reminiscences of their native places, as well as trying
to adjust the new environment to their previous way of life, as may be observed
in the houses of *Celsa*, Ampurias, *Bilbilis* or Caminreal. This practice can be
observed all over the territories conquered by the Romans.

It seems appropriate to finish by citing another text of Martial that reflects the
idealization of a remote place. Although he felt happy in his indigenous home in
Bilbilis, he could not forget the delights of Rome:

XII, *Epist.* to *Priscus*
[…] I miss that audience of my fellow-citizens to which I had grown accustomed,
and seem to myself a pleader in a strange court; for whatever is popular in my

small books my hearer inspired. That subtlety of judgment, that inspiration of the subject, the libraries, theatres, meeting-places, where pleasure is a student without knowing it – to sun up all, those things which fastidiously I desert I regret, like one desolate. Added to this is the back-biting of many fellow-townsmen, and envy ousting judgment, and one or other evilly disposed persons – a host in a tiny place – [...].

Both Roman and indigenous people felt nostalgia, not only for their homeland, but also for the places that had made them happy. Martial is a perfect example that illustrates the complexity of human nature one presumes transcending all historical periods.

References

Balil, A. 1971. Casa y urbanismo en la España antigua. *Boletín del Seminario de Estudios de Arte y Arqueología* 37, 311–28.

Balil, A. 1972. Casa y urbanismo en la España antigua. *Boletín del Seminario de Estudios de Arte y Arqueología* 38, 55–131.

Balil, A. 1973. Casa y urbanismo en la España antigua. *Boletín del Seminario de Estudios de Arte y Arqueología* 39, 115–88.

Barra, M. 1996. La Casa in Magna Grecia, in *Richerche sulla casa in Magna Grecia e in Sicilia, Archeologia e Storia* 5, 41–66.

Beltrán, M. 1991. La Colonia Celsa. *Actas de la casa urbana hispanorromana.* Zaragoza: Institución Fernando El Católico, 131–64.

Beltrán, M. 2003. La casa hispanorromana. Modelos. *Bolskan* 20, 13–63.

Beltrán, M., Mostalac, A. and Lasheras, J.A. 1984. *Colonia Victrix Iulia Lepida-Celsa. Zaragoza. I: Arquitectura de la Casa de los Delfines.* Zaragoza: Institución Fernando El Católico.

Beltrán, M. and Mostalac, A. 1994. *Colonia Victrix Iulia Lepida-Celsa* II: *Estratigrafía y pinturas.* Zaragoza: Institución Fernando El Católico.

Beltrán, M., Minguez, J.A. and Lasheras, J.A. 1998. *Colonia Victrix Iulia Lepida-Celsa* III: *Instrumentum.* Zaragoza: Institución Fernando El Católico.

Bonini, P. 2006. *La Casa nella Grecia romana. Forme e funzioni dello spazio privato fra I e VI secolo. Antenor-Quaderni 6.* Roma: Edizioni Quasar.

Brunneau, P. 1972. *Exploration archéologique de Délos. Les mosaïques.* Paris: De Boccard.

Bruno, J. y Scott, T.R. 1993. *Cosa IV. The Houses.* Memoirs of the American Academy of Rome, Vol. 38. Rome: The University of Michigan Press.

Coarelli, F. 1970–71. Classe dirigente romana e arti figurative. *Dialoghi di Archeologia* 4–5, 241–65.

Coarelli, F. 1983. Architettura sacra e architettura privata nella tarda Repubblica. *Collection de l'École Française de Rome* 66, 191–217.

Fernández Vega, P.A. 1993. *Arquitectura y urbanística en la ciudad romana de Julióbriga.* Santander: Universidad de Cantabria.

Garcia-Entero, V. 2005. *Los balnea domésticos -ámbito rural y urbano- en la Hispania Romana. Anejos de AEspA* XXXVII. Madrid: CSIC.

Gros, P. 2001. *L'Architecture romaine II: Maisons, palais et tombeaux.* Paris: Picard.

Ling, R. 1991. *Roman Painting.* Cambridge: Cambridge University Press.

Martín-Bueno, M. and Saenz, C. 2001–2002. La insula I de Bilbilis. *Salduie* I, 127–58.

Martín-Bueno, M; Sáenz, C; Reklaityte, I. y Uribe, P. 2007. Baños y letrinas en el mundo romano: El *balneum* de la *Domus* I, Insula de las Termas, Municipium Augustam Bilbilis (Calatayud) *Zephyrus* 60, 205–23.

Rawson, B. and Weaver, P.R.C. 1997. *The Roman Family: Status, Sentiment and Space.* Oxford: Oxford University Press.

Russo, A. 1996. Le abitazioni degli indigeni: problematiche generali. in *Richerche sulla casa in Magna Grecia e in Sicilia, Archeologia e Storia* 5, 67–87.

Saller, R.P. 1984. Familia, Domus and the Roman Conception of the Family. *Phoenix* 38, 336–55.

Santos, M. 1991. Distribución y evolución de la vivienda urbana tardorrepublicana y altoimperial en Ampurias, *Actas de la casa urbana hispanorromana.* Zaragoza. Institución Fernando El Católico, 19–34.

Scagliarini, D. 1983. L' Edilizia residénciale nelle cittá romane dell'Emilia Romagna. *Studi sulla città antica. L'Emilia-Romagna.* Roma: Erma di Bretschneider, 283–334.

Treggiari, S. 1999. The Upper-class House as Symbol and Focus of Emotion in Cicero. *Journal of Roman Archaeology* 12, 33–56.

Uribe, P. 2004. Arquitectura doméstica en Bilbilis, la Domus I. *Salduie* 4, 191–220.

Vicente, J., Martín, R., Herce A.I., Escriche, C. y Punter, P. 1989. Un pavimento de opus signinum con epígrafe ibérico. *Mosaicos romanos in Memoriam Manuel Fernández-Galiano*, Zaragoza, Institución Fernando El Católico, 11–42.

Vicente, J., Punter, Mª Pª, Escriche, C. y Herce, A. 1989. El mosaico romano con inscripción ibérica de la "La Caridad" Caminreal, Teruel. *Xiloca* 3, 9–29.

Vicente, J., Punter, Mª Pª, Escriche, C. y Herce, A. 1991. La Caridad (Caminreal, Teruel). *Actas de la casa urbana hispanorromana*, Zaragoza, Institución Fernando El Católico, 81–130.

Wallace-Hadrill, A. 1988. The Social Structure of the Roman House. *Papers of the British School at Rome* LVL, 43–97.

Wallace-Hadrill, A. 1990. The Social Spread of Roman Luxury: Sampling Pompeii and Herculaneum. *Papers of the British School at Rome* LVIII, 145–93.

Wallace-Hadrill, A. 1991. The Houses and Households: Sampling Pompeii and Herculaneum. In B. Rawson (ed.), *Marriage, Divorce and Children in Ancient Rome.* New York: Routledge 191–229.

Wallace-Hadrill, A. 1994. *Houses and Society in Pompeii and Herculaneum.* New Jersey: Princeton University Press.

Wallace-Hadrill, A. 1997. Rethinking the Roman Atrium House. In R. Laurence and A. Wallace-Hadrill (eds), *Domestic Space in the Roman World: Pompeii and Beyond.* Portsmouth: *Journal of Roman Archaeology* 219–40.

Wiseman, T.P. 1997. *Conspicui postes tectaque digna deo*: The Public Image of Aristocratic and Imperial Houses in the Late Republic and Early Empire. *Collection de l'École Française de Rome* 98, 393–413.

Chapter 10
A Scent of Plaice?

Antony Firth

Introduction

When archaeologists think about the importance of places, it is implicit that there is a material environment within which the relics of past human activity can be perceived physically, usually visually. Humps, bumps, buildings, vegetation, vistas – whether recognized or not – instil a sense of history; a sense of place emerges from a physical, visual environment. A past that has meaning in the present can be perceived directly from one's surroundings.

Figure 10.1   The site of the 70-gun warship *Restoration*, lost in the Great Storm of 1703, looking from the Goodwin Sands across the historic anchorage of The Downs towards Richborough in Kent

Source: © Wessex Archaeology

At sea there are no fixed surroundings. There are no physical or visual prompts. The visible surface of the sea embodies no trace of its past. Pressed by the weather, its form alters by the second. Does this mean that, without an evident past, the sea can generate no sense of place, only an anonymous ever-changing present?

This is surely not the case. The sea in Figure 10.1 lies over the wreck of the *Restoration* in Goodwin Sands, off Kent. To the observer this landscape might appear to have been wiped clean by the sea, but to marine archaeologists and others familiar with the centuries of maritime activity in this space it is as rich as any palimpsest on land. Hence specific bits of sea can have rich cultural and historical meaning amongst the public both locally and at large, and not only when framed by a coastline (Gibbs 2005).

My own early experiences of archaeology underwater were as a volunteer diver in the Solent in the mid-late 1980s having no formal archaeological training – just my own interest and the fantastic input of the people running the Isle of Wight Maritime Heritage Project. The sense of place I obtained from these early experiences could be very strong. On the wreck of the sixteenth-century merchant ship *Santa Lucia* (Watson and Gale 1990) I was struck by being the first person for 400 years to be amongst the timbers, pottery and pewter I was excavating. Crossing on the ferry from Yarmouth to Lymington, the coastal saltmarsh evoked the earlier estuary represented by the Holocene sedimentary sequence we had been investigating at Bouldnor Cliff, now 5–15m below the water I was crossing. I've spent a lot of time since on marine sites and looking at the coast and sea to contemplate the ways in which marine and maritime seascapes have been used and experienced in the past. I've also shared the sense of place I get from knowing a little about the sea's histories in the course of explaining what we are doing and why to developers, boat crews and passers by. Whereas my early career was based on being on and in the sea, I'm more reliant now on other people's investigations and experiences, obtaining a sense of place indirectly from a report, computer screen or meeting in order to help interpret and advise on what elements of the historic environment might be present or significant.

My work today is informed by research into the relationship between archaeologists and society at large, including the way that archaeologists intervene in the physical environment and people's perception of it (Firth 2002). That research provides a basis for exploring how 'a sense of place' works at sea despite a lack of immediately visible prompts. Existing public sensitivity to marine places may be an important building block in extending and deepening public understanding of the past and the role of the sea within it. Actively seeking to extend such public sensitivities may also be important in implementing effective heritage policies in such an openly-accessible environment. Understanding a sense of place at sea might also help in understanding a sense of place on land, especially where the visual evidence is ephemeral.

Mechanisms

Drawing from earlier work, the mechanisms whereby a sense of place can arise at sea are explored here by combining the theoretical approaches of three authors: the sociologist Anthony Giddens; psychologist J.J. Gibson (1904–79) and philosopher of science Michael Polanyi (1891–1976) (Firth 2002).

There are three ideas from Giddens' work that are particularly relevant to understanding sense of place: structuration; locale; and the stratification model of consciousness. Giddens' theory of structuration is that the apparently fixed things in society – 'structure' – only exist in the innumerable instances of individual activity that make up everyday life. Each human action is enabled and constrained by the effects of previous actions, and goes on to enable and constrain subsequent actions. Structure does not have a separate existence, therefore, but is being repeatedly remade. This constant re-making is clearly emphasized by Giddens: 'Every act which contributes to the reproduction of structure is also an act of production, a novel enterprise, and as such may initiate change by altering that structure at the same time as it reproduces it' (Giddens 1976 [1993]: 134).

Giddens maintains that people's physical surroundings are incorporated into the instances of human activity as 'locale'. In effect, the physical world is integral to structuration, and is also engaged in the constant process of being re-made by human actions (Giddens 1984: 118–19). Locale shapes human activity, and is simultaneously shaped by it. Again a novel enterprise, the repeated re-making of the physical world is driven in part by how people create locale at the point of action.

Giddens also suggests that human consciousness is stratified, referring to two types of consciousness, practical and discursive. In Giddens' conception, huge amounts of daily activity are conducted through practical consciousness – thoughtful and reasoned but not subject to explicit consideration, which is the realm of discursive consciousness (Giddens 1984: 5–7). A key aspect of Giddens' stratification model of consciousness is that thinking can move between the layers: practical consciousness can be subject to discursive reflection; discursive thought can subside into practical consciousness.

The psychologist Gibson developed a school of thought termed ecological psychology, arguing for 'direct perception', which is a kind of subliminal understanding of the world that arises directly from being within it. In contrast to the notion that the eye forms images on the retina that are then subject to visual processing, Gibson argued that the attributes of the environment are understood directly on the basis of what the environment gives to – 'affords' – the inhabitants of the environment: 'the basic properties of the environment that make an affordance are specified in the structure of ambient light, and hence the affordance itself is specified in ambient light' (Gibson 1979 [1986]: 143). Observers are, therefore, integrated perceptual systems: 'between observation and judgement lies no shadow of thought' (Fenton 1993: 44).

Direct perception offers a powerful account of how a sense of place might work. Old things in the environment impose history and meaning upon our perceptions directly. They don't require processing; they are understood intrinsically.

However, direct perception appears not to help in understanding a sense of place at sea, where there is an absence of features that can be perceived. Gibson's view also appears universal and environmentally deterministic. It does not sit well with Giddens' idea that we are constructively engaged with an environment that we make and re-make.

Polanyi suggested that all knowing draws on tacit knowledge, characterized as 'knowing more than we can tell' (Polanyi 1967: 4). People necessarily make assumptions in order to find out something new, so all understanding is based on working from a proximal (near) term to a distal (distant) term (Polanyi 1967: 10). This distal term becomes a new proximal term as personal knowledge addresses the next distal term. Consequently, people extend themselves into the world through their knowledge, progressively inhabiting a new tool or idea as a basis for extending themselves towards the next.

Polanyi regards the individual's body as the irreducible proximal term: 'our body is the ultimate instrument of all our external knowledge' (Polanyi 1967: 15). Tacit understanding of proximal terms is therefore likened to dwelling; people dwell in their unspoken knowledge in order to understand the world. As distal terms become understood they are interiorized, forming the tacit, proximal term from which a new distal term is approached: 'we keep expanding our body into the world' (Polanyi 1967: 29).

Polanyi's account of tacit knowledge sits well with Giddens' stratified consciousness. We engage creatively with the world by progressively extending ourselves into it, but at each extension the distal term that was once approached discursively now subsides into practical consciousness as a tacit, proximal term that is no longer remarked. The meaning of the distal term comes to be known practically as we dwell within it.

The once distal, now proximal terms within which we dwell can be perceived directly in the manner identified by Gibson. In the instances of daily activity we observe and judge the environment within which we dwell without any shadow of (discursive) thought. The difference from Gibson is that direct perception has not arisen automatically from our physical surroundings, but has been acquired creatively by progressively expanding into the world. A sense of place does not emerge from places mystically, or like some strange radiation. Rather, a sense of place is brought to that place by the people who inhabit it, though the original creativity has been interiorized to such a degree that it is unrecognized.

If this approach is valid, it implies that sense of place arises because we can perceive a meaningful past directly from our physical surroundings, but this meaning is not inherent in those surroundings. Rather, meaning arises from what we – in the present – have attributed to these things, but which have now subsided into practical consciousness. In Polanyi's terms, we dwell in all sorts of subliminal ideas about the past that are evoked by our surroundings through Gibsonian direct perception, and these ideas about the past shape (and are re-shaped by) our day-to-day activity through Giddens' structuration. Importantly, because meaning arises not from physical surroundings but from what we have interiorized about those surroundings, we need not – after Gibson – perceive only that the sea is wet, wobbly and not easily walked upon; if we have come to dwell in its history, the sea can afford a sense of place.

Implications

Dwelling, in Polanyi's sense, is all the more interesting because it does not necessarily imply physical dwelling. One analogy that he uses is a probe. Initially, an individual may concentrate on the sensations of the probe in the hand. These sensations become tacitly accepted as the individual comes to dwell in the probe as an extension of the body, with their attention focused on the end of the probe as they prod and poke. Other writers drawing on Polanyi have spoken of dwelling in other kinds of tools, including electronic instruments, such as echosounders:

> … experienced skippers often speak of knowing the details and the patterns of the 'landscape' of the sea bottom 'as well as their fingers'. This indicates that for the skilled skipper fishing technology – the boat, electronic equipment and fishing gear – is not to be regarded as an 'external' mediator between his person and the environment but rather as a bodily extension in quite a literal sense. (Pálsson 1994: 910)

This is important because so much archaeological work at sea is dependent on electronic tools for sensing the physical environment. If Polanyi's view is accepted, then dwelling and perception is no less authentic for being mediated electronically. We can expand into the underwater world using instruments, remote sensing and assumptions about the effects of marine processes upon archaeological material. Equally, a sense of place can be afforded to people who can never visit the seabed through geophysical imagery, video, podcasts, computer reconstructions or models. This conception also allows that I can perceive a sense of place at sea from a desk, computer or meeting, and not only when I have salt on my lips. It also reduces a bias in landscape archaeology towards physical and especially visual perception, allowing more space for perceiving the world through other senses and providing, thereby, a useful corrective in considering the archaeological experiences of all sorts of people whose access or visual sense is impaired.

Another important implication of the notion that sense of place is something acquired through an active, creative process is that it draws attention to the 'how' as much as the 'what'. What is sensed from a place depends on how an individual has proceeded from proximal to distal term, and how those distal terms have since been interiorized. Consequently, there is a multiplicity of paths through which people come to acquire direct perception of a sense of place. So although the place may be the same, the sense of it may be multifarious. Furthermore, the learning experience through which one group of people – such as archaeologists – may have acquired a sense of place may be quite different to how other groups have come to dwell there; sense of place need not be shared. If we come to knowledge through different paths we may find differences – even conflicts – in perception. Such conflicts may be all the more intractable because they are buried subliminally in practical consciousness.

If the key to sense of place is not *what* we know, but *how* we know, then for archaeologists to communicate about our sense of place with others – children; politicians; local stakeholders – we have to invite them to dwell not in *what* we can see in the historic environment, but in *how* we see. Recognizing that the process of knowing is reducible ultimately to our own bodies, simply telling the public details of our technologies and techniques – the probes we use – is insufficient; we have to share the fundamental assumptions of archaeological interpretation (Firth 2002: 29).

Understanding the mechanisms through which people come to know a place may indicate how, by sharing our assumptions, archaeologists can enhance other people's sense of place. Equally, insensitivity to these mechanisms may result in archaeologists inadvertently causing damage to people's sense of place. In rendering practical dwelling into discursive consciousness, the displacement of an inherent sense of a place with an explicit identification of meaning can be experienced as a loss (Firth 2002: 25–6). When archaeologists share their understanding of places with the public, careful attention is needed to avoid circumstances where 'the fascination of discovery ... is ... replaced by the tedium of display' (Barrett 1995: 6).

The possible damage to people's sense of place may be all the more serious because of the relationship between sense of place and identity. Gibson notes that 'to perceive the environment is to co-perceive oneself' (Gibson 1979 [1986]: 141) whilst Giddens' emphasis on people's maintenance of their own biographical narratives indicates that each individual's day-to-day interaction with their environment is constitutive of self (Firth 1992: 17). Thoughtless archaeological intervention in sense of place, even if well-meaning, may cut to the core of identities that people tacitly inhabit. In a parallel relating to social reformers and planners seeking to help inhabitants of industrial housing escape from the grimness of their surroundings, Harevan and Langenbach quote Marris: 'They identify with the neighbourhood: it is part of them, and to hear it condemned as a slum is a condemnation of themselves too' Marris, P. (1974) *Loss and Change*, p. 55 (quoted in Harevan and Langenbach 1981: 115). Nor is sense of place only a question of meaning; referring to Giddens again, it is important to acknowledge that legitimacy and power are also – simultaneous with meaning – incorporated within people's sense of place (Firth 2002). A sense of place may also be a sense of knowing one's place.

Considering these implications, it can be seen that where, for example, we are at the position at which a certain vessel sank, the sea may afford a sense of place and we may come to perceive the history and tragedy from the waves themselves. The sense of place need be no less if the wreck itself is only visible as a blip on an instrument; or if, for example, the commemorative dropping of wreaths on the water is being watched on television. And amongst the viewers, that sense may be quite different according to their experience; if it is off their shore, or on a route they have traversed, or the ship was built in a local yard, or their family tree has members lost similarly, and so on. Archaeologists' dry factual accounts of remains on the seabed may not – in the minds of some – 'do justice' to those lost; and a sophisticated explanation of exploratory technology might only emphasize professional disinterest. At the same time, archaeologists might equally be dwelling in a glorious maritime past through a sense of place that is evoked subliminally by physical prompts such as pub signs and cannon-shaped bollards far inland. Undoubtedly important to British history and identity, Trafalgar is a well-known place. But how many people have been there, or actually know where it is?

Application to Submerged Prehistoric Places

On the face of it, the case for a sense of place at sea might most easily be made for places that were once land, namely areas of prehistoric landscape that have been submerged as a result of sea-level rise. Knowledge of submerged prehistoric landscapes has been current among a relatively small circle of archaeologists and quaternary scientists for decades (Coles 1998; Gaffney et al. 2009) but there seems to have been a tangible increase in uptake and acceptance in the last decade or so amongst marine developers and the public at large. Undoubtedly, Coles' embodiment of the concept in the term 'Doggerland' (Coles 1998) has played a large part in the emergence of a sense of place. 'Doggerland' has been incorporated into the title of archaeological books and journal articles (e.g. Ward et al. 2006; Gaffney et al. 2007; Gaffney et al. 2009), in popular archaeological media including television and radio, and has spread beyond archaeology to entitle music albums, photography, poetry and a North Sea surfing blog.[1] The success of the term arises,

---

1   For example:

Television:	http://natgeotv.com/uk/stone-age-atlantis;
	http://natgeotv.com/uk/stone-age-atlantis/galleries/doggerland/1;
	http://www.channel4.com/history/microsites/T/timeteam/2007_
	dogger.html.
Radio:	http://www.bbc.co.uk/programmes/b00lmpkb.
Music:	http://www.doggerland.com/.
Poetry:	http://postcardsfromdoggerland.wordpress.com/about/.
Photography:	http://www.lightstalkers.org/galleries/contact_sheet/4692.
Surfing:	http://thedoggerlandchronicles.blogspot.com/.

it seems, because Dogger-land, as a distal term, can be approached from familiar proximal terms: Eng-land; Scot-land; Nether-lands and so on. Clearly – and thankfully – this was Coles' intention: to shift discussion of submerged landscapes in the North Sea away from landbridges to being an inhabited land whose influence on cultural activity – specifically to understanding the Mesolithic-Neolithic transition – has to be considered in its own right. Beyond this – 'a landscape ... here called Doggerland to emphasize its availability for settlement' (Coles 1998: abstract) – the constitution of Doggerland is rarely discussed. Although the term Doggerland might encourage people to start sensing a lost country whose borders can be mapped and its community imagined, direct evidence for this new old land is still scant. Modern nation-states – England, Scotland, Netherlands – are an inappropriate proximal term from which to approach the distal evidence. If a sense of place is perceived directly from the name Doggerland, whether by archaeologists or the public, then discursive attention is required not only to the evidence but also to the influence on our approaches to submerged prehistory of models of territory and social organization drawn from modern society.

In Coles' papers (1998, 2000) Doggerland is bounded in time and space to the Mesolithic and to the (Southern) North Sea. However, the term is also being used to encompass evidence for submerged landscapes over wider periods and geographical extents. The fragmentary character of the data is well known to researchers, but Doggerland serves as a useful conceptual container to hold the bits together. This handy container might, however, subside into practical consciousness as an unremarked, proximal assumption that – even if remaining distal to the researchers themselves – bestows an unwarranted sense of place among the public. If this is the case, then Doggerland is achieving a sense of place more certain than the disparate data allow, extending all around Britain and enduring tens and hundreds of millennia.

Whilst there is no question that prehistoric deposits and archaeological material have been identified at places like Bouldnor (Momber 2000) and Area 240, off Great Yarmouth (Tizzard et al. in press), and the body of evidence is growing rapidly, the presence and survival of these elements has been subject to specific local circumstances that reflect large-scale glacial and cultural processes over very large spans of time. Even 'landscape' implies rather more sense of place than the evidence often deserves. Far from the cosy notion of Doggerland comprising a single, continuous landscape fringing the UK and Continent whilst the sea – rising after the Last Glacial Maximum – laps against its shores, submerged prehistory around the UK encompasses a 700,000 year history that was frequently devastating and has left large areas stripped to bare rock or comprising tens of metres of jumbled glacial debris. It is possible to pick out places where fragments of former landscapes have survived, but these places have to be understood first in their own terms – deposits separated by millennia may lie stacked one upon another – and be only cautiously linked to neighbouring places or across their region.

Although they must be kept distal and discursive rather than proximal, a sense of these places should not be denied entirely. It is important to offer, as Coles

Figure 10.2   Image from a computer-generated animation of the Mesolithic River Arun, 11 miles off Sussex and about 25 metres below today's sea-level, based directly upon the results of geophysical, geoarchaeological and palaeo-environmental investigations

Source: © Wessex Archaeology

and others have, inferences that integrate topographic, palaeo-environmental and archaeological data, and to postulate the remnants of prehistoric landsurfaces offshore as inhabited places (see Figure 10.2). It is also important to share these interpretations with the public who fund the work, and whose appreciation of the richness of the marine environment and the enormous changes wrought by climate change is to be encouraged. However, for submerged prehistory a sense of place has to be carefully circumscribed in time as well as place, whilst archaeologists share with the public the highly uncertain state of current archaeological knowledge.

Application to the Wrecks of Ships and Aircraft

In contrast to relict landsurfaces, the capacity for wreck sites to furnish a sense of place is quite different to monuments of comparable age and size on land. Wrecks under the sea often lack a form comparable to their form when in use. That is to say, monuments on land – even if barely standing or reduced to sub-surface features – usually retain their original plan and relation to the topography of the surrounding landscape. On the one hand there are at least some commonalities between an observer in the present and the original inhabitants, providing a proximal base for a sense of the place to be perceived directly. On the other hand, wrecks of both ships and aircraft are often heavily disrupted, crashing to the seabed on their sides, upside

**Figure 10.3   The steam engine of a wreck in the English Channel thought
to be the Belgian steamer *Concha*, sunk by a collision in 1897.
The engines, boilers and fittings have a monumental character
that evoke a sense of place equivalent to industrial sites on land**

Source: © Wessex Archaeology

down, or broken; and further deformed by sedimentary processes or clearance for
navigation, for example. The craft that now lie wrecked were also never intended to
be experienced on the seabed; they are alien to the environment in which they are
now observed, and that environment is itself alien to most observers. Consequently,
there is an absence of familiar proximal terms in which people might dwell. For
many people, wrecks are unfamiliar forms in an unfamiliar environment whose
otherness requires that they be attended discursively as distal terms, impeding the
development of a sense of place that might be perceived directly.

Despite the cataclysm of loss, some wrecks retain considerable coherence.
Photographs, video and geophysical images can be readily reconciled with the
original ship or aircraft. Such sites are often valued for the quality of survival,
but might a favourable predisposition to coherent sites arise because a sense of
place emerges more readily from these familiar forms than from significant but
shattered remains? Similarly, emphasis on wrecks that have associations with
famous individuals might be explained in terms of a sense of place arising from
tacit familiarity with these historical characters.

**Figure 10.4   The propellor and engine of a B-17 Flying Fortress in the
English Channel. The aircraft is unidentified so its remains
present a tantalising connection to crew, mission and
circumstances of loss**

Source: © Wessex Archaeology

Increasing availability of imagery of wrecks as they actually are will help
people to interiorize the form and environment within which seabed wrecks lie,
allowing them to approach new distal terms and giving greater scope for a sense
of place to emerge. But it should be noted that a sense of place from a wreck does
not only arise from the location of the wreck itself. Again different from most
monuments on land, ships and aircraft appear to have sense of place associated
with their being 'movable places'. So as well as the wreck site, there may be
a sense of the ship or aircraft in terms of its place on its 'normal' or last route
– the air or sea not far removed from its current location. Similarly, a wreck site
may invoke a sense of the ship or aircraft as a place itself, in which people lived,
worked and died.

The sense of place of a wreck site is also – at least in part – a sense of event;
in this case, a sense of the event of wrecking. This prompts the thought that in
many cases – both on land and at sea – what we experience as a sense of place
is really a sense of the events that have happened at that place. On land this may
be indistinguishable or immaterial, because both 'place' and 'taking place' are

co-located. But for a moveable item, the sense of place associated with a wreck might also encompass events in which the ship or aircraft took part, but at a different place. For example, the wreck site of U-boat *U86*, scuttled off the Isle of Wight in 1921, also evokes its role in a war crime south west of Ireland in 1918, when its Captain ordered the sinking of the hospital ship *Llandovery Castle* and the massacre of its survivors (Wessex Archaeology 2007: 22–4). Comparable examples are provided by ships in preservation or in museum contexts; *Victory* invokes a sense of Trafalgar and the death of Nelson – events rather than places – despite being in a dry dock in Portsmouth. In contrast, a considerable sense of place arises from the *SS Great Britain* being in Bristol, in the actual dock in which its revolutionary construction took place.

Conclusions

These few examples start to illustrate how the sense of place arising from prehistoric landsurfaces, wrecks and other maritime monuments can be explored. There are clearly many more facets to a sense of place at sea that warrant analysis but I hope I have established, at least, that sense of place can arise even if there are no direct visual prompts. This is because sense of place is not an intrinsic characteristic of the places themselves; rather, it is a product of what the observer brings to that place. The contribution that people make to places that gives rise to sense of place may not be explicitly known to the people themselves. As I have argued, sense of place is perceived directly as a result of people dwelling in knowledge that is already familiar. People can become familiar with the presuppositions of marine archaeology or maritime history, for example, and enjoy a sense of marine places as a consequence. Nor does sense of place depend on being in immediate contact with the place itself; as we dwell in the technologies of investigation, we can sense the places we explore.

This account highlighted the importance of the details of the process through which sense of place arises, noting that differences in the process can result in differences in sense of place. The discrete processes applying to archaeologists and non-archaeologists can result in places being differently perceived at a fundamental level; and I have also indicated that in rendering the basis of people's experience of a place explicit, archaeologists may damage the sense of place those people enjoyed. Sense of place is, therefore, also an arena in which archaeologists must consider their responsibilities. Archaeologists can seek to measure, record and assess sense of place; they might even seek to increase awareness – providing snippets of knowledge and understanding that can be assimilated by the public, to provide them with proximal terms from which sense of place can develop. What are the responsibilities of archaeologists as they intervene more proactively in sense of place? Should archaeologists enhance sense of place, or even start to manufacture a sense of place where it appears lacking? The process described here is certainly amenable to instrumental manipulation, but are archaeologists entitled to operate within people's practical consciousness, where interventions have to be 'under the radar' to be effective? And given that openness may impair sense of place, does its manipulation – however well intentioned – run counter to scientific obligations regarding explicit, transparent application of data and methods?

Archaeologists have a privileged relation to the past, having spent years extending themselves into ever more complex techniques, theories and datasets and able to dwell, therefore, in a particularly rich understanding of old places. Archaeologists also have the habit of questioning, subjecting proximal terms to discursive scrutiny – with the challenge of new questions perhaps offsetting the loss from explication. One answer to concern about archaeologists' responsibility towards sense of place might be to worry less about improving 'theirs' and concentrate more on sharing 'ours' with a wider public.

References

Barrett, J.C. 1995. *Some Challenges in Contemporary Archaeology.* Oxford: Oxbow.

Coles, B.J. 1998. Doggerland: A Speculative Survey. *Proceedings of the Prehistoric Society* 64, 45–81.

Coles, B.J. 2000. Doggerland: The Cultural Dynamics of a Shifting Coastline. In K. Pye and J.R.L. Allen (eds), *Coastal and Estuarine Environments: Sedimentology, Geomorphology and Geoarchaeology.* London: Geological Society.

Fenton, P.C. 1993. The Navigator as Natural Historian. *Mariner's Mirror* 79 (1), 44–57.

Firth, A. 2002. *Managing Archaeology Underwater: A Theoretical, Historical and Comparative Perspective on Society and its Submerged Past.* BAR International Series 1055. Oxford: Archaeopress.

Gaffney, V., Fitch, S. and Smith, D. 2009. *Europe's Lost World: The Rediscovery of Doggerland.* CBA Research Report 160. York: Council for British Archaeology.

Gaffney, V., Thomson, K. and Fitch, S. 2007. *Mapping Doggerland: The Mesolithic Landscapes of the Southern North Sea.* Oxford: Archaeopress.

Gibbs, M. 2005. Watery Graves: When Ships Become Places. In J. Lydon and T. Ireland (eds), *Object Lessons.* Melborne: Australian Scholarly Press, 50–70.

Gibson, J.J. 1979 [1986]. *The Ecological Approach to Visual Perception.* Hillsdale, New Jersey: Lawrence Erlbaum Associates.

Giddens, A. 1976 [1993]. *New Rules of Sociological Method: A Positive Critique of Interpretative Sociologies.* Cambridge: Polity Press.

Giddens, A. 1984. *The Constitution of Society: Outline of the Theory of Structuration.* Cambridge: Polity Press.

Hareven, T.K. and Langenbach, R. 1981. Living Places, Work Places and Historical Identity. In D. Lowenthal and M. Binney (eds), *Our Past Before Us? Why Do We Save It?* London: Temple Smith, 109–23.

Momber, G. 2000. Drowned and Deserted: A Submerged Prehistoric Landscape in the Solent, England. *International Journal of Nautical Archaeology* 29 (1), 86–99.

Palsson, G. 1994. Enskilment at Sea. *Man* 29 (4): 901–27.

Polanyi, M. 1967. *The Tacit Dimension*. London: Routledge and Kegan Paul.

Tizzard, L., Baggaley, P.A. and Firth, A.J. Forthcoming. Seabed Prehistory: Investigating Palaeo-landsurfaces Associated with a Palaeolithic Tool Find, North Sea. In J. Benjamin, C. Bonsall, C. Pickard and A. Fischer (eds), *Underwater Archaeology and the Submerged Prehistory of Europe*. Oxford: Oxbow.

Ward, I., Larcombe, P. and Lillie, M. 2006. The Dating of Doggerland: Post-glacial Geochronology of the Southern North Sea. *Environmental Archaeology* 11 (2), 207–18.

Watson, K. and Gale, A. 1990. Site Evaluation for Marine Sites and Monuments Records: Yarmouth Roads Wreck Investigations. *International Journal of Nautical Archaeology* 19 (3), 183–92.

Wessex Archaeology 2007. Wrecks on the Seabed R2: Assessment, Evaluation and Recording. Appendix C: Archaeological Results. Unpublished Report for English Heritage. ALSF Project Number 3877. WA Ref: 57454.03(a). Available at http://ads.ahds.ac.uk/catalogue/adsdata/arch-473-1/ahds/dissemination/pdf/Project_Reports/Text/Round_2/2006/57454_Appendix_C_final.pdf.

Chapter 11

Sense and Sensitivity –
Or Archaeology Versus the 'Wow Factor' in Southampton (England)

Duncan Brown

The aim of this chapter is to examine the notion of 'sense of place' through the ways in which a place relates to and promotes its own past. 'Heritage' is often seen as something that imbues a town, district or region with a 'sense of identity' that somewhat dispiritingly is often boiled down to a promotional tool for the garnering of visitors. British roads for instance, are now bedecked with signs welcoming travellers to heritage high spots such as 'Robin Hood country'. Such is the desperation of some places to find a marketable identity that they even promote themselves as the settings for television fictions. This search for a 'selling point' encapsulated in a slogan is superficial of course, because every place, however that term may be defined, is too complex to be reduced to a single sentence. Indeed, the search for something unique in a place is ultimately self-deluding. If we believed every bit of tourist sloganeering then we would presume that the whole country is surrounded by impressive beaches, every corner of the countryside has something to marvel at and every community offers the warmest welcome you could ever wish to receive. At such a level, there must surely be very little difference between for example, one sweeping range of hills and another.

The connection between a particular town and an Elizabethan/Chaucerian/ Austen-esque (*etcetera, ad infinitum*) story will not therefore be apparent just by walking the streets, whilst logos and taglines derived from one event or character do not really put across any sense of what a place is about. A true understanding of the past provides context, offering a deeper perspective on places, communities and buildings that allows similarities, as much as differences, to be understood and celebrated. The question considered here is how much, if at all, the study of the past, or more precisely archaeology, informs a sense of identity in a particular place and how archaeological interpretations enter the public consciousness. This exercise will concentrate on one particular community, where the author worked as an archaeologist for 26 years, 14 of them in the capacity of curator of the Museum of Archaeology.

Southampton

The port city of Southampton has found several ways of positioning itself in the competitive heritage marketplace. Citizens and visitors alike may follow the Jane Austen trail for instance, which attempts to make the most of the famous authoress' rather less well-known residence in the town between 1806 and 1809. The Spitfire fighter aeroplane was developed and built in Southampton and a recent campaign has been lobbying for the erection of a Spitfire memorial. Looming much larger is the story of the *Titanic*, which sailed from Southampton with many local people as crew. The death-toll was 1,523 and 549 of those victims came from Southampton families. The city contains many memorials, the most spectacular being that to the engineers, that now comprise a *Titanic* tour. These are of course fleeting incidents in a 2,000 year settlement history that began with the coming of the Romans.

Archaeology provides glimpses of the prehistoric past but there is little evidence for nucleation prior to the Claudian invasion and most archaeological research has concentrated on the three known Roman, Saxon and medieval towns. There remain few tangible traces of the earliest two of those but substantial stretches of the medieval town wall have survived, within which may be found many

Figure 11.1   The Bargate, the principal gateway into the medieval town

medieval vaulted cellars and several medieval houses and churches. Southampton has the third best preserved circuit of medieval town wall in England, the oldest purpose-built artillery fortification in the country (God's House Tower, currently the Museum of Archaeology) and the Bargate, an imposing medieval gateway (Figure 11.1). There is thus much that is noteworthy beyond the stories and personalities that customarily have the highest profiles. Southampton may also justifiably claim to be amongst the birthplaces of urban archaeology in England, with a series of excavations in the 1960s, including those published by Colin Platt (1975), informing developments in other towns and cities. There has been archaeological excavation almost continuously since the late 1940s and the results therefore should by now be an established source of information and inspiration for those planning the future of the city.

If there is a link between understanding the past and establishing a sense of place then getting a feel for the public perception of archaeology in Southampton might help to chart how that sense has developed, at least since the end of the Second World War. That is actually a good starting point for this exercise because the town was devastated by enemy bombing and subsequently has become a very different place from that which existed before 1940. This may perhaps be most pointedly illustrated by the following comment that appeared in an independent feasibility study carried out in 2003 into the development of a new museum in the city: '*there is no obvious central focus in the city – it lacks a heart/identity*'.

Southampton is and has always been a port and for much of its history it seems almost the whole community was involved in port activities. The story of the town (only recently a city) must therefore be rooted in its relationship with the water. Jane Austen and the *Titanic* are mere reflections of that greater theme, for Austen came to live with her sea-captain brother, while the *Titanic* was one of many liners in Southampton's long history as a centre for oceanic travel. It is easy to fasten onto iconic figures or stories in order to build up a marketable profile but that does not fully capture the essence of the place. Southampton and its past are more profoundly represented by the people who lived and worked there; groups of merchants, seafarers, dock workers, shipbuilders and more.

The earliest parts of that continuous story can most easily be reached archaeologically, and nearly all the excavated evidence relates to the functioning of a port community. This is reflected in the museum of archaeology, which displays Roman, Saxon and medieval finds relating to trade across the English Channel with France and beyond. Many of those exhibits are wondrous objects in their own right, worthy of use in promoting Southampton's 'heritage'. If that does not happen it may be because they have no obvious relevance to a particular story. The story of archaeological investigation within the city boundary is worth visiting here, however, as an indicator of the part it has played in giving Southampton a sense of place. At the same time it is interesting to look at the ways in which the presentation of that archaeology to a wider public has fluctuated. One way of doing that is to review how archaeology and museums have appeared in the pages of the *Southern Daily Echo* (*The Echo*), Southampton's local newspaper. Press cuttings reflect what was instantly significant within communities and this approach should therefore give some insight into the story of Southampton's heritage and how it might have contributed to a sense of place.

Museums

Table 11.1 shows a selection of headlines relating to Southampton's museums and ancient buildings while Table 11.2 is more specifically related to archaeological

Table 11.1   Headlines from the *Daily Echo* relating to museums in Southampton

Year	Headline/story
1942	Blast Reveals Treasures 'Echo' War Museum suggestion noted archaeologist approves
1948	Too few exhibits chasing too few museums
1951	Bargate Museum draws 6,700 in first five days
1959	Archaeologists praise care of town's medieval structures
1960	Piece of Soton history comes to life … as museum
1962	Museum of Science and Industry plan
1965	Big plans for its future but … The Old Warehouse Still Stands Empty
1966	£100,000 museum group for Southampton
1966	Maritime museum is new Southampton showpiece
1976	Museum hours to be cut
1987	Maritime Centre to be 'launched'
1989	Southampton's … maritime heritage centre … will not be ready until at least late 1991
1995	Museum charges to be dropped Attendance fall forces u-turn
2000	£60,000 probe into plans for heritage centre
2002	Heritage crumbling away
2002	A multi-million pound heritage centre commemorating the Titanic disaster is today exclusively revealed
2003	The WOW factor Southampton is to get a 4.6 million pound heritage centre based around the sinking of the Titanic
2005	A giant museum of Southampton – shaped like a Spitfire wing – is unlikely to be built in the city
2008	'Visitors will have a genuine experience rather than it being just a glorified museum' – Councillor John Hannides

discoveries. The year 1942 is a good point at which to begin this brief survey, because the town had suffered so badly in the Blitz of 1940 that it must have been very difficult for the surviving inhabitants to sustain any sense of belonging. Post-war rebuilding programmes did not resurrect what had been destroyed and the town that grew out of the rubble was nothing compared to what it had been. At such times a community needs to re-discover its identity and that was the focus of much of the early press coverage relating to museums. *The Echo* promoted the idea that museums, as purveyors of the past, could re-establish Southampton's identity. The 1942 story, where the paper called for a museum to be built out of the rubble of the war, was initiated by the exposure during the bombing of archaeological remains. O.G.S. Crawford, who had already recovered evidence for Saxon Hamwic from bomb-sites, was the noted archaeologist who lent his support to the story. It was *The Echo* that called for the opening of a new museum and refused to let the idea go, for in 1948 they ran a story about the paucity of gallery space in the town. The interest of both these pieces is that they reflect a local desire to re-discover Southampton's roots, and the extent of that desire is reflected in the headline from 1951 ('Bargate Museum draws 6,700 in first five days').

No museum in Southampton now attracts visitors in those numbers over such a short period. The Bargate Museum, opened inside the imposing medieval gateway in the northern town wall, presented a fairly basic story of the development of Southampton but there was obviously a huge desire, a need even, among the local community, to interact with and be informed by the past. Such high levels of interest make it easy for town councillors to take decisions in support of such initiatives and the next two headlines show how that continued into the 1960s. In 1959 the Society for Medieval Archaeology held its second annual conference in Southampton and *The Echo* ran a story about how impressed those conference-goers were with the way the town's medieval buildings were being preserved and cared for ('Archaeologists praise care …'). Accompanying the piece was a photograph of various luminaries of the Society standing before the Bargate. It was also revealed that the Town Council was spending £30,000 in that year on the 'restoration and conservation of the old walls and other structures'; no small amount in those post-war years.

One of those 'other structures' was God's House Tower which was being renovated from a roofless shell to become the archaeology museum. That opened in 1960, to be greeted by an appropriate headline ('Piece of Soton history …'), the thrust of the story being that this was a much-needed development, at last providing a permanent home for the wealth of archaeological artefacts being excavated from bomb-sites within the medieval town. Reference was made to the opinion that until now it had been 'quite possible to come into the port and see nothing of 1,000 years of history'. This is another important aspect in the development of museums in Southampton. Not only were they a means by which the community could re-establish its roots; they could also stand as a representation to visitors of a town's past, thus becoming a source of local pride.

It did not take long, however, for a familiar pattern to develop. Plans to turn Eagle Warehouse, a fine example (dated to 1903) of the sort of building that was once common in pre-war Southampton, into a museum of science and industry were shelved, then dropped, despite the attentions of *The Echo*. The 1962 story ('Museum of Science and Industry Plan') presented the idea for the new museum as a logical extension to the programme of development that had introduced the archaeology museum and was creating a maritime museum. In 1965, however, *The Echo* ran a story ('The old warehouse still stands empty') highlighting the lack of progress with the museum of science and industry and Eagle Warehouse remained unoccupied until 1983, when it became a store, as it still is, for the City Council's archaeology collection. The Maritime Museum opened in 1966, inside the medieval Wool House, and was trumpeted as part of a grand vision ('£100,000 museum group for Southampton') that would bring the city five specialized museums. These included the Archaeology and Maritime museums, Tudor House Museum of Local History and the Bargate. Although the museum of science and industry was never added to that quartet, this remains the period of greatest investment in the museums of Southampton.

Ten years later the headlines announced cuts in opening hours that ultimately made the Museum of Archaeology accessible on only one Saturday a month, while the others were closed every Monday. That remained the case until the mid-1980s, when two new permanent archaeology galleries on the Saxon and Medieval towns were installed, with somewhat less press interest. Focus shifted instead to plans for a new maritime museum located in the newly developed Ocean Village marina that was supposed to include floating exhibits (1987: 'Maritime centre to be "launched"'). Those plans developed to the point where two working boats were acquired as centre-pieces of the new museum but delays were announced in 1989 and soon after the whole project was dropped completely. A new cinema was built on the site instead.

Southampton's museums thereafter appeared fleetingly in the press until, at the behest of the City Council, admission charges were introduced in the early 1990s. This resulted in a two-thirds decline in visitor figures, forcing a 'U-turn', as *The Echo* put it in 1995. At the turn of the century plans for a new museum, or heritage centre as Council members and managers preferred to call it, were back in the headlines (2000: '£60,000 probe …'). This time £60,000 was put into a feasibility study that identified a potential site and set out possible themes and exhibits. At the time this was known as 'The Story of Southampton' and it was to combine the archaeology, local and maritime museums into one site offering state-of-the-art galleries and displays. The Council agreed a further sum of money to follow up the feasibility study with a scoping project. That project was never commenced and press attention focused instead on the lack of investment in existing structures such as Tudor House, which was closed to the public and in a state of disrepair (2002: 'Heritage crumbling away'), some contrast with the headline in 1959. Tudor House has now been saved by a substantial lottery-funded project that has seen the building renovated and new displays are being planned in advance of opening in 2011. This is a rare success story, for other monuments are increasingly at risk as budget cuts deepen and newly re-introduced museum admission charges have again reduced visitor figures.

The lure of a new museum remains however, and no sooner was 'The Story of Southampton' shelved than 'Sea City' appeared (2002: 'A multi-million pound heritage centre'). The centrepiece of this new development, located in the now defunct magistrate's courts in the western block of the Civic Centre is apparently to be a huge replica version of part of the *Titanic*, designed to provide visitors with the 'genuine experience' which, in the opinion of the present Member for Leisure, will make this more than 'just a glorified museum' (2008). The main ingredient, of course, must be the WOW factor, first introduced in 2003 as the thing to draw the crowds and maximize Southampton's 'heritage offer'. The mere concept of a 'Titanic experience' indicates that the City Council would place sensation over sensitivity and commercialism over community. This development, at least, is more realistic and tasteful than the suggested giant Spitfire wing (2005) with which Southampton's independent Heritage Federation courted the headlines. The Heritage Federation is made up of heritage groups that are independent of Southampton City Council and plays the part well of a useful, if unpredictable, thron in the corporate side.

Nobody doubts the need for a new museum. God's House Tower and the Wool House are medieval buildings that, in terms of access and environment, are no longer suitable as museums. It is difficult to install lifts to the galleries and objects are at risk from the lack of atmospheric control. The buildings themselves are also suffering, as the masonry walls dry out and spall under the effects of central heating. The 'Sea City' project has now received a grant of £4.6 million from the Heritage Lottery Fund and Southampton is closer to providing its citizens with a new museum than it has been for decades. Another £10.1 million of funding has yet to be found, however, while the target opening date of 2012 is approaching all too rapidly. This project, like so many others in Southampton, could yet go the way of its totemic ocean liner and sink into oblivion.

This necessarily brief review of the development of Southampton's museums brings out a number of issues around the idea of place. One such issue is the belief that Southampton's identity is best reflected in a single theme. For some that is the Spitfire, for others it might be Jane Austen but in terms of saleability the *Titanic* is regarded as the biggest marketing tool. Although 'Sea City' is intended to be more like a museum than the coverage allows, press releases have concentrated on the *Titanic* theme, perhaps because it is believed to be the story most likely to gain the project a national and international profile.

There is, of course, more to Southampton than a single maritime disaster but it is possible to observe no recognition, among senior figures in the City Council, of the many stories explored within the existing museums. It may also be true that the hopes of Southampton's post-war residents have been betrayed, for the town has not quite re-discovered its identity. Some indeed apparently still seek that identity in the period of the Second World War itself, preferring to focus on the stories of the Spitfire and the Blitz rather than the place Southampton was before then and has become since.

A second issue is how the press cuttings reveal a gradual decline in the notion that the past is irrevocably bound up with a sense of local identity. During and after the Second World War there seems to have been genuine enthusiasm for the development of museums as a way of providing a new sense of communal purpose and as a reflection of Southampton's history, achievements, tragedies and personalities. The past then did seem to offer a context for the present. It is difficult now however, to envisage press coverage of a Society for Medieval Archaeology conference in Southampton. Since the 1960s the newspapers have focused on stories that frequently seem to set the Council up for a fall, telling as they do of declining investment and failed plans. The museums have been built but are now regarded as out of date and not 'fit for purpose'. It is to the credit of the Council that they plan to develop a new facility but a 'visitor experience' does not sound like something that will foster, or even represent adequately, any sense of Southampton's identity.

Archaeology

Table 11.2 is a selection of headlines relating to archaeological activity in Southampton. The theme here is one of continuing interest in major discoveries, from the turtle bone found in the mid-Saxon town (1951) to the gold from Saxon graves at the new football stadium (2001), although Norman pottery (1957) and cobblestones (1958) would probably not now be considered newsworthy. Another consistent thread seems to be the search for a catchy headline such as 'Medieval Southampton had Atmosphere' (1957) and 'Clausentum's kebab culture' (1996). In between however, the headlines reveal how archaeology has changed in the city. In 1953 the visit of the Mayor to an excavation was considered newsworthy but now it is unlikely that he or she would even consider such a thing and if they did it would probably not be reported. In 1963, after the Museum of Archaeology had opened, Southampton's archaeologists were 'heroic'. Working to reveal Southampton's past was considered to be an important part of post-war regeneration, even though those who did so battled against insufficient resources and poor conditions. In 1971 the founding of the Southampton Archaeological Research Committee (SARC) was recorded, as the amount of excavation reached new levels, but problems with developers in 1974 ('Firms refuse to contribute') presaged greater pressures to come. A year later the director of SARC left Southampton for Dorset, telling *The Echo* that this was 'due in no small way to the fact that the Council are not prepared to commit themselves to archaeology in Southampton' (1975: 'Just scratching the surface').

The City did eventually set up its own archaeological field unit, which still exists (1980: 'City's past does have a future'; 1981: 'The past is ours in future') and SARC ceased its applications for funding from the Department of the

Table 11.2   Headlines from the *Daily Echo* relating to archaeology in Southampton

Year	Headline/story
1950	1,229 years ago Hamwic was a trading centre Hampshire Field Club visit excavations
1951	The turtle of Hamwic is a mystery
1953	Mayor sees excavations
1957	Norman pottery discovered in latest Soton 'dig'
1957	Medieval Southampton had atmosphere! A 300-year-old stench is hampering archaeologists…
1958	Southampton's Romans used cobble-stones
1960	Archaeology now appeals even to the young
1963	Heroic archaeologists in the 'goon-like world' south of Bargate
1969	Fourth year of the big dig
1971	'Digs' new Soton group is formed
1974	Firms refuse to contribute
1975	Old relics are deteriorating
1975	'Just scratching the surface'
1980	City's past does have a future
1981	The past is ours in future
1986	Hectic year for archaeologists
1991	The future of Southampton's history is under threat
1992	Archaeology powers used
1996	Clausentum's kebab culture
2001	They've struck gold under Saints stadium

Environment. The press coverage then returned to covering excavations (1986) as the new structure became established and the Manpower Services Scheme provided a large workforce. In 1991 however, the effects of economic recession were recorded as 'secret talks look set to end in the city council's archaeological department being cut back to a skeleton staff'. A year later the introduction of PPG16 was described as 'new planning rules mean that the city council can insist on archaeological evaluation being carried out before any site is built on' (1992). The theme underlying those stories is that the people of Southampton value archaeology as an integral part of understanding the city's past. One question arising from that is how that understanding informs the future of the city.

The Museum of Archaeology in Southampton was developed as a direct result of research, the continuation of which has informed subsequent displays. Such research can therefore deepen and enrich a sense of identity, although it may not always excite those looking for catchy one-liners to entice new audiences. It is one thing to develop a new museum, or 'visitor experience' that may or may not incorporate the results of research, while it is another to refer to the past in the way the city as a whole is planned and developed. The streetscape can become an immediate way of referring to the past as a means of reinforcing a sense of place.

Development

The North foyer of the Civic Centre in Southampton currently houses an un-textured model of the centre of the city, which presents a plan for future developments under the title 'City Vision Southampton' (Figure 11.2). This is designed to act as a focus for discussion, 'a place where industry professionals can meet, where local school children can learn about the built environment and where people can discover how the city is to change over the coming years'. Certain points on the model are numbered in reference to wall panels that describe completed or impending '*key developments*' such as a new Scandinavian flat-pack furniture emporium. The credibility of the exercise is tested, to say the least, by attendant statements such as: 'the city is what it is because our citizens are what they are'; and 'a city is a language to the building of which every citizen brings a stone'. Superficial slogans

Figure 11.2   City Vision: an installation in the Civic Centre, Southampton

such as these lead one to question the integrity of this project, for one suspects the influence of marketing practice. This feeling is scarcely assuaged by closer inspection. Plans for a development within the medieval walled town, entitled 'The French Quarter', are described as 'carefully grafting a new piece of city fabric to reintroduce the historic street pattern and create a vibrant new residential and commercial heart'. The theme is maintained in the 'Old Town Development Strategy', which can be found on the city council website. The 'Old Town Vision' explains that:

> the Old Town's historic character was its enclosed, intimate, and built-up nature, resulting from its geography, events and medieval fortifications [and that] this pre-war character should be re-established through re-instating the historic street pattern where possible, the built form's mass and scale, and encouraging the use of appropriate high quality materials. [It is also stated that] the Old Town's historic buildings and archaeology should be interpreted to help reinforce and promote its distinct identity.

Figure 11.3   A view of High Street to the North

Figure 11.4   A view of High Street to the South

A walk around the 'French Quarter' and High Street areas of Southampton's Old Town does not altogether confirm the accuracy of those claims. Figures 11.3, 11.4 and 11.5 are views from the point where the French Quarter development meets High Street. Figure 11.3 looks up High Street to the North, where surviving Victorian buildings, a medieval church and a timber-framed pub sit among later structures filling gaps created by bombing. Figure 11.4 shows twenty-first-century developments along the South of High Street that do not seem to echo the intention to re-establish the pre-war character. Looking West in Figure 11.5, up a re-instated historic thoroughfare towards a medieval merchant's house restored by English Heritage, it is difficult to see how the new French Quarter development is a successful manifestation of either historic mass and scale, or the encouragement of the 'use of appropriate high quality materials'. More importantly there is little evidence of a distinct identity informed by historic buildings and archaeology. That might be achieved by a genuine recognition of mass and scale and actually using high quality materials but another way of referring to the past in new developments is to provide information about the site.

Figure 11.6 is the first in a series of tile panels shown on the side of a building erected in the 1980s on the site of Southampton Friary and this is a simple, if somewhat superficial, method that can be found in many other towns and cities. There are no such information panels among the new buildings in the French

Figure 11.5   A view from High Street through the new buildings of the 'French Quarter' development towards the medieval merchant's house

Figure 11.6   One of a series of tile panels incorporated into a 1980s development on the site of Southampton Friary

Quarter despite the extensive and productive excavations that took place there. These developments have altered the streetscape to the extent that older inhabitants find it unrecognizable as familiar elements are swamped by larger buildings. The aim of course is not to construct an environment with which existing residents can identify but to create something that will bring in new 'customers'. Marketing strategies are thus developed to emphasize history but there is no obvious understanding of what that really means, in terms of context, continuity and a true sense of place.

Conclusion

Reviewing the ways in which archaeology has been represented in the press shows how important the past has been, and still is, to the people of Southampton. The attention of the press is aroused more often now when major discoveries are made, or when a jokey headline can be constructed and if it seems that archaeology is no longer as important to journalists as it was, then that possibly reflects the way in which archaeology has become a mainstream activity within the development process. Press coverage of the museums suggests an early purpose and determination, when museums seem to have been regarded as a way of re-discovering a sense of identity, followed by a continuing cycle of projects started and shelved. Despite that, there remains a creditable ambition to develop a new museum that will properly showcase the city and it's past.

It is also apparent that the museums, and the attendant activity of archaeology, have yet to be fully integrated into a coherent conception of the present and future of the city. A 'visitor experience' smacks more of a retail outlet than the truly involved and involving encounter that a museum can be. Once again, although the emphasis has been on *Titanic* as a selling point, it is the application of research-based knowledge and interpretation of a continuous past that is most likely to allow a Southampton museum to transmit a true sense of place. That may not, of course, be the function of this new attraction, in which case the aspirations of the post-war community will have been failed. If however it is intended to create displays that reflect aspects of Southampton's past then it would be good to reflect as much on what makes Southampton the same as what makes it unusual.

Archaeology is often the study of the mundane or a comparison of similar sets of evidence in order to fit a place into a wider pattern. We would understand nothing if all our findings were unique. We are thus returned to the introduction, where it is suggested that aspects of similarity can be important in the search for a sense of identity. The past shows us with what we can identify and that is useful to know when the character of a community needs to be established, as was the case in the 1940s. More and more, however, it seems that despite claims to the contrary, the city is losing sight of that identity as it develops. It seems ironic that planners are promoting individuality just as every high street in England is lost to the same selection of chain stores. The past, therefore, has lost none of its potential for informing communal development but we seem to have lost the willingness to appreciate that. The close links between archaeology and the planning process ought to provide the means of rediscovering the importance of identity and the true meaning of heritage. Instead, archaeology is too often viewed as an unwelcome expense rather than a vital exploration of the human context. That will not change until archaeologists challenge society to recognize the value of what they do and, as the press cuttings show, that has been achieved before.

References

Platt, C. and Coleman-Smith, R. 1975. *Excavations in Medieval Southampton, 1953–1969*, 2 vols. Leicester: Leicester University Press.
Southampton City Council. 2004. 'Old Town Development Strategy'. Available at http://www.southampton.gov.uk/Images/7%20Old%20Town%20Vision_tcm46-161135.pdf. [accessed: 2 March 2009].

Chapter 12

Ilhna Beltin: Locating Identity in a
Fortified Mediterranean City

Rachel Radmilli

This chapter is based primarily on a short ethnographic film that was produced for the EuroMed Heritage II project 'Mediterranean Voices'. The film – *Ilhna Beltin* – or Voices of Valletta – reflects the same aims and intentions as the project itself which include, among others, giving a voice to individuals and groups that inhabit Valletta, listening to their stories and examining the ways they define their sense(s) of place. An extract from the description of the film highlights the concept of communication and use of space where tradition and modernity bleed into each other within a small yet monumental city: 'The documentary film aims to highlight the everyday, the mundane, the "hidden" and the possibly less glamorous, popularized or gentrified aspects of life in Valletta today and within living memory.'

Through discussion of a few aspects of this film and the impact it has had on different stakeholders or audiences the chapter will also explore the contribution such media can make on managing and understanding intangible heritage, as well as other forms of heritage that fall within the framework of this publication. Can we come to a true understanding of senses of place and what can we learn from the experiences of *Ilhna Beltin* to better inform policy-making that relates broadly speaking to the management of people's lives, landscape, identity and sense of belonging? And given a chance, how would the different stakeholders themselves go about defining their sense of place, if such a thing exists? These questions have already been answered by different people involved in this project and the subtleties evident in their critique are reflected in the discussion below. Furthermore, people's reaction to the film can also help us understand the very same debate that underpins these questions.

This film – *Ilhna Beltin* – has been shown in a variety of contexts from the class-room to film festivals and conferences. This chapter itself is a follow-up to the European Association of Archaeologists (EAA) conference session held in Malta in September 2008. The conveners of this session defined 'sense of place' as being, 'commonly used in professional and domestic situations to describe the emotional attachment people have to the places they hold dear. This sense of place – sometimes referred to as "*genius loci*" – can equate with what has been termed "the lure of the local", with its concern for the familiar – the place where we live, or where we lived when we were children. It is also about rootedness, belonging, stability and identity.' This same broad definition is adopted here as it refers closely to different groups and individuals who had something to do with this film either as interviewers or interviewees. The argument below will therefore develop by first discussing the way the film fits within the profile of the Mediterranean Voices project. Secondly it will consider how people define their sense of place and identity by invoking a sense of rootedness and belonging where the imposing walls of this fortified city offered refuge as well as myriad opportunities to be explored at different times of their lives. The next part of this chapter will also take a look at the impact this film has had on other people's lives and the way we also define our sense of place, not only in relation to Valletta but also to the way we think about Valletta and our own sense of identity. I will be describing some personal experiences of Slavko, the film maker/cameraman who happened to be living in Valletta by choice and not by birth. A Serbian national by birth, Slavko experiences a transformation during the making of this film where he finally gets accepted by his neigbours, making him feel part of the city and of the neighbourhood. Different groups of university students have also seen the film and their reactions are interesting and worthy of discussion. Not only are they the best and most honest critics, but some comments can help us come to a clearer understanding of what sense of place actually means.

One point that emerges is that sense of place is not just metaphorical in the way de Certeau (1988: 115) outlines it but also about literally using all the senses: to *see* things clearly albeit subjectively, to *smell* the surroundings – where smell is often the best way to invoke a memory, and to *hear* – anything from the church bells marking time, to listening to people's stories, opinions and experiences. These will be the three main senses I will explore while assessing the experiences of the three groups of people mentioned here.

Project Background

This film was produced as one of the outputs by the Maltese team for the Mediterranean Voices project. This project involved a consortium of 13 different partners from different parts of Europe and the Mediterranean and was coordinated by London Metropolitan University. The main focus was to study intangible heritage within an urban context and included the whole notion of changing political and physical landscapes, oral histories, identity and changing use of space. The urban lifestyle was clearly framed within a number of historic cities as diverse as Las Palmas, London (where numerous migrants from the Mediterranean have moved to create a new life), Valletta (a city that has lost most of its original population to several demographic changes), Alexandria (with its glamorous cosmopolitan past), Beirut, Ancona, Nicosia (a divided city), Marseilles (a crossroads between North and South), Bethlehem (a political island), Istanbul, Chania, Granada and Ciutat de Mallorca.

Each partner in the consortium took different aspects and realities pertinent to the place in question as their main focus, but the shared histories and common focus brought us all together as one team. To quote Selwyn (2005: 205):

> ... at the heart of the Med-Voices project, in both its theoretical outlook and its practice, lies the Mediterranean as imagined – physically, metaphorically, and digitally – by our researchers and their interlocutors. [Scott 2005] describes the conception, shaping and experiencing of the project in terms of imagined and lived communities that are not so much constrained by concrete state and other borders, policed as these tend to be by armed border guards, walls and surveillance towers, but that are shaped and integrated by imaginative recognition of regional continuities and cosmopolitan coexistence.

The digital dimension played an important role in this project in that one of the aims was to produce a database and a website giving researchers, students, scholars and a wider audience the opportunity to explore this virtual Mediterranean that is made up of so many different and personal stories. So much movement leads to overlap and one common factor is that boundaries are always blurred. Migration is a common theme in the history of the Mediterranean region so it is normal to hear stories of the Maltese in Alexandria or of families of mixed heritage. These issues always raised further questions such as how one defines one's identity as French, Egyptian, Maltese, Palestinian or Spanish? Beyond that are the personal and ethical questions to be asked, especially when dealing with digital media.

One researcher on the project from Chania in Crete talks about his experience taking photos for his team. Varvantakis observes how public spaces, 'can be easily photographed. But things become more complicated when the lens turns to face the individual, for new possibilities and dangers arise. What happens when one gets inside an individual's house – or when one gets to interact with people, to listen to their stories and perceive an amount of their personalities and histories? How could or should we photograph all this?' (cited in Selwyn 2005: 208). So bearing in mind the definition of sense of place cited above, certain terms become blurred. Can we really talk about rootedness when so many people are constantly on the move? Can we talk about locality when people share such mixed heritage? The closer we look at the particular, the harder our task becomes. Yet this is no reason to abandon the exploration of such diversity. This sense of place and identity can and often is quite fluid and rhythmic, as Bachelard (1994) tells us, where we can see that by moving from the general to the particular or from the grand to the mundane we can appreciate the complex nature of people's perception and sense of identity associated with place and time.

When talking about sense of place we can forget that researchers of this subject face several dilemmas and ethical issues which in turn have a very personal effect on their lives. This in a way is one of the reasons why I have chosen to include the experience of the film maker in the discussion of this chapter as the experience has had a very real impact on his life and his sense of belonging to Valletta –

something that takes us beyond senses of place to consider one's sense of identity and how one fits into a place as an 'outsider'. There are many discourses to be reconciled here and many issues to be negotiated. Negotiating identities and the different layers of narrative is always a tricky question – all the more so in the contemporary Mediterranean when so many different realities coexist with one another. Yet each city has a different story to tell.

Valletta has a very particular story. It is a fortified Mediterranean city that is possibly the smallest city covered by the Mediterranean Voices project. It is a city with a very specific origin and particular military design and history. Built in 1565 by the Knights of the Order of St John, it was built on a grid plan to become the new capital city of the Maltese Islands, and is now inscribed as a UNESCO World Heritage city. It is built on a promontory with two natural harbours, one on either side – the Grand Harbour and the Marsamxetto Harbour. Until recently the main access into the city was from the sea-side of the fortifications where trade, military activity and all sorts of maritime activity dominated the lifestyle and identity of Valletta. It is only after the Second World War that the orientation of the city changed to face inland. The Grand Harbour played an important role in the Second World War as the Mediterranean Fleet was positioned there. The many ship and plane wrecks now lying on the seabed are testimony to this. Sadly, so were the heavily bombed neighbourhoods that were located closest to the harbour where many families had to migrate out of their homes and temporarily move to the other side of Valletta, or out of the city altogether. This temporary migration often took on a permanent form after the war when several families never moved back to 'their' part of Valletta.

Following the war much migration also took place out of the country where many Maltese families moved to places such as Australia and the UK in search of work and a better lifestyle in the 1950s and 1960s. Malta eventually gained independence in 1964 after which businesses related directly or indirectly to the services dwindled. New movements took over – both social movements and political movements to 'clean up' the place. So the red light district in Strait Street was closed down; other neighbourhoods were demolished and the original population uprooted completely for political reasons by the then prime minister. Consequently the displaced communities and the associated 'business' moved to different parts of the island, and to Soho in London. Over time the overall population of Valletta shrunk from circa 18,000 in 1957 to circa 7,000 in 1995 (MEPA 2002), a situation that is proving difficult to reverse. As a result, Valletta becomes a ghost-town after closing hours. Although it is a commercial centre (that is now also facing competition from other towns) and a place where government and legal offices are located, thousands of Maltese and tourists visit during the day, but leave again at night (Figure 12.1). Even the hotels in the area are somewhat derelict and the nightlife is sparse. This is in stark contrast with most – or all – capital cities around Europe and the Mediterranean.

Over the past five years, several initiatives have been considered to try and reverse this situation, and many are becoming popular. In summer events are

Figure 12.1   Pjazza Regina, Valletta
Source: Photo: Rachel Radmilli

organized under the combined themes of the Malta Arts Festival, and a *Notte Magica* is staged on a regular basis where restaurants, shops and bars remain open all night and several performances – including street performances – are organized for the event. The cruise liner terminal has been developed and enhanced to breathe new life into the harbour area and an area by the coast known as the Valletta Waterfront has been restored and renovated turning the historic stores first built by the Knights into an entertainment area and functional promenade. The problem here though is that these events and sites attract numerous but temporary visitors, who then leave to go home to different locations around the island. The status quo remains.

Ilhna Beltin

The film, *Ilhna Beltin* (2005) was created with these different realities in mind as well as with ideas of identity, heritage and communication. Although Valletta has a low population density, the sense of identity is nevertheless extremely strong. Even those who left Valletta still feel strongly about their heritage and their affection for, affiliation with, and loyalty to the city. These feelings are often demonstrated

Figure 12.2   Statue of St Paul, Valletta
Source: Photo: Rachel Radmilli

and expressed in clear contexts such as at football or waterpolo matches as well
as during the various parish feasts where political competition and affiliation
emerges clearly where two major feasts are associated with specific political party

supporters in a way reminiscent of Blok's (2001) 'minor differences' argument (Figure 12.2). These were some of the issues that we tried to explore in the film, where every street has meaning and where everything is so contained in the small city, but yet so rich and complex. The film tried to play on the different layers of life and the different groups who pass each other by and sometimes intermingle.

The Protagonists

The film was based primarily on interviews with two people born in Valletta and who have lived all their lives there. One woman speaks about her experience working in one of the famous bars in Strait Street after returning from a short stint in Australia after the war. Her husband was out of work at the time of their return to Malta, and they had four children to bring up. She tells how the city itself has a life and draws on the identity of people living there. There is an incredibly strong attachment with the place and she states categorically that she will never again leave Valletta – leaving Valletta for Australia, albeit temporarily, sounds like the one big regret in her life. The only way they can take her out, she says in the film, is when she dies. This statement works because there is no cemetery located within the fortifications of the city, and in fact she died, shortly after the film was produced.

The second personality is a well known jazz musician. His interview reminds us that in the clean-up process, when there was a move to redress the city's sleazy reputation, a lot was at stake and was lost or sacrificed. Many musicians launched successful careers in the bars of Strait Street – if they made it there, they could literally make it anywhere. And he tells us that wherever you went in Valletta, you knew who you would meet and where certain people would hang out. People were a part of each neighbourhood and those neighbourhoods had their own identity, an identity often associated with social class, political affiliation, and often with particular professions. Much of this local knowledge is being lost with the emigration of much of the original population and gentrification taking place in the city, as well as the influx of previously unrelated new residents who risk outnumbering the older families. Most of the families living in Valletta had a history and strong attachment to the place. Their heritage was located there and their families were all born and raised there for generations. There was a sense of acceptance, recognition and understanding, where histories were shared and the sense of community was strong before this significant population decline took place.

Both interviewees speak with a strong sense of nostalgia for times past, notwithstanding the fact that they were tough times economically and politically. Yet Valletta was still a vibrant city at the time and this was reflected in the sense of identity and memories of people we worked with on the project. Although most of the physical damage was repaired, the musician described how Valletta was 'finished' or 'over'. It is no longer what it used to be. So this is where the irony lies: that people's sense of identity is still so strong. They really want something to be done about the situation – and ideas are forthcoming.

The Film Maker

A similar point was made about sense of belonging and the way it would relate to identity, and acceptance. This resonates well with the story of one of the film makers. Slavko is originally from Serbia. He rented a flat in Valletta and tried hard to become part of the community, only to have his efforts compromised by a friend who, while visiting from Belgrade, caused a rift with the neighbours. This next quote is a long one but is a speech that Slavko made at the launch of the film for the film festival we organized for the project on 3 February 2006. In his own words, broken English and humorous style he states:

> My name is Slavko Vukanovic ... I have [made] a documentary called *Voices of Valletta*. I was an interviewer and a camera assistant. As my name suggests, I am not Maltese, but a Serbian and I have been living in Malta for a period of seven years. Living in a country as small as this one for seven years makes you part of a community, makes you local, makes you hear and know things you didn't really need to know ...
>
> ... I reached an agreement with Marie-Claire about renting her place in Carmelite Street in Valletta. I moved in there in November 2003. [The] place looked lovely: arches in the bedroom, very cool furniture, bit dark, no sunlight whatsoever and ghosts in the bathroom but still a very unique place; I liked it very much. I had a balcony with a view to another balcony. [For six] months of the year both of us open our balcony doors and its like I got a living room extension into their living room. Lina and her husband, who lived opposite my place, were a kind couple. They were updating me with information that I didn't really need. Something like that Maria the *hoxxna* is carrying someone else's child or whatever. ... However, shortly after I moved in, my friend Ali came to Malta, to my wonderful place, to chill out from the Belgrade ghetto. The first weeks of his [visit] were turbulent; he was drunk every night, playing loud Serbian hip hop, shouting and celebrating life. Seven policemen came to my house at 11 o'clock one evening responding to a report from my neighbours. We were not arrested by a miracle. He was offering cookies to the police, insisting that they had to try them no matter what they asked him. Somehow, the police left my place that night without even filing a report or taking the cookies.
>
> The neighbors hated us after that. Ali went back to Serbia a month later. The neighbours hated me after his departure. Months passed by with only our local undertaker and his family smiling to me. They got a garage just round the corner of my house, where amongst the coffins they keep a 42 inch TV they watch all day long.
>
> Then Edward Said came along with the idea for shooting *Voices of Valletta*. My street seemed like a perfect location to describe the balcony communication

environment of Valletta. My neighbours were the main protagonists of the documentary and they all loved it. They all got paid for it. Then they fell in love with me. And they kept updating me with information I didn't need to know.

We shall return to Slavko later.

Students

The third user-group I wish to discuss involves the students who were shown this film in different classes and at different times. We had students in Malta reading for a BA in Tourism Studies who followed a course on the project. I had showed it to some other students in Morocco while working on a Tempus programme (and we also showed them some of the other films produced by the project), and it was shown to a group of archaeology students as part of several MA courses on offer at the University of Bristol. Their comments and criticisms were interesting and relevant to this discussion.

One group was quite critical of the technique adopted where certain scenes were staged to recreate the memories described by the interviewees. The sense of nostalgia was played out by actors in authentic costumes recreating the old bars in Strait Street frequented primarily by the sailors working for the services during the war. Interestingly enough the archaeology students felt that this was a kind of distraction that deterred from the story itself. The students felt that the acting as well as the use of subtitles detracted from real life. More space should have been given to the interview with more focus on the ethnography, implying that there was no room for poetic license. This is a valid point in the light of the present discussion. How can a discussion of sense of place aid policy-makers if at all? This film served a number of purposes – recording aspects of personal life, the intangible heritage was presented in tangible format, and the film itself was used for didactic purposes. Yet the approach and reactions could teach us a lesson as to how to deal with intangible heritage broadly speaking, and senses of place more specifically. This film was seen as a kind of testimonial to real people's lives.

This same sentiment was reflected in the reaction by the Moroccan students. As a slight divergence here, I wish to mention their reaction to a film produced by our colleagues in Bethlehem that dealt with marriage customs among different religious groups within living memory, with a focus on present-day wedding ceremonies. When I asked the students for their comments and reactions, I was taken aback. The subject of the film itself was irrelevant to them. What was surprising to them is that it made them realize something so obvious, but which has been obliterated from people's minds. For them it was almost disturbing to realize that notwithstanding the news headlines constantly referring to the peace process, the Israel-Palestine conflict, wars, terrorism, segregation, migration etc, that what we often forget is the most obvious thing – that real people are living real lives there. People have a life, and are trying to live it no matter what. People still fall in love, get married, argue with their mothers about the style of their wedding dress, still offer the token dinar, and such wonderfully mundane everyday concerns that affect people's lives. And this ironically made the conflict all the more difficult to accept. A dose of reality, or of the mundane, a little bit of real people's life stories was all it took to make this dramatic point.

This same point is reinforced by Slavko's words. And this possibly is what policy-makers need to keep in mind. For policy-making to work we cannot just *think* of the people. We cannot just focus on the official, or on the big news, or on the major headlines. We cannot just empathize with local populations or think about what 'they' need. It is no use imagining what it would be like to try and understand them. Empathy may take us one step further, but Slavko's comments really hit home: it was by *working with* them that his relationship was restored with his neighbours. True, he mentioned that they were paid for their work, but they were paid a token fee, so that is possibly irrelevant. But they were the stars of the show. They were involved, and they actively participated in the production of the film about their city. And Slavko then really became a part of the community: '*wiehed minn taghna*' which means he became 'one of "us"'.

Policy-making

We are constantly living with images and using images to communicate with each other. Furthermore, digital media and communication are contributing to a globalized lifestyle. Szerszynski and Urry (2006: 113) actually talk about the way television has led to people becoming virtual travellers and to people taking on an imagined cosmopolitan lifestyle. They argue that, 'the changing role that visuality has played on citizenship throughout history also involves a transformation of vision and absenting from particular contexts and interests ... we then draw on place and vision to argue that the shift to a cosmopolitan relationship with place means that humans increasingly inhabit their world at a distance'. This adds another dimension to ethnographic film in that it can take on an ambassadorial role. As in the case of the Bethlehem film, while many are scared to visit Palestine/ Israel given the current state of affairs, the film acted as an eye-opener to many audiences who could finally come closer to life as experienced in a particular place. Can these different journeys, or emergent cosmopolitanism as the authors call it, have significant social, cultural and political implications? My only difficulty here is with the concept of 'emergent' in relation to the Mediterranean since as a region it has a long history of cosmopolitanism and population movements for trade, migration etc. The Mediterranean Voices project highlights these same implications albeit for fear of a loss or dispersal of this diversity. So on the one hand digital media have been used to record and store oral histories and personal experiences, while also reaching out to audiences in other parts of the world who would otherwise not have access to these same places.

Selwyn (2005: 216) confirms that the project highlights different contexts and interrelationships: 'Social and cultural continuities are in the process of being undermined by a regional (and global) politico-economic system … [several examples] inevitably lead to questions about the relationships between power, profit and the displacement of the weak: contexts, surely, for political expressions of cultural difference.'

Bianchi (2005) also states that some of the aims of the project include:

> [promoting] an awareness of the cosmopolitan histories and multi-ethnic cultural practices which contribute to the intangible cultural heritage of Mediterranean urban landscapes, and, to disseminate this information as widely as possible amongst as wide an audience as possible: academics, policy-makers, civil society organizations, educational institutions, local residents, and amongst members of dispersed Mediterranean populations living and working outside the region. [Futhermore the project also aims to] pursue a commitment to the social and cultural pluralism of the Mediterranean and to contribute to the nurturing and maintenance of intercultural dialogue, tolerance, and cooperation amongst the different individuals, groups and communities who share these urban spaces. (2005: 294)

This emphasizes the idea of active cooperation between so many different stakeholders, from residents to policy-makers to migrant communities who need to actively cooperate with each other for the benefit of these localities or urban spaces. Slavko spoke humorously about the ghosts in his bathroom, however the image invoked is why so many of these people's identities are rooted so deeply in this urban space. First- and second-generation migrants still feel strongly about their homeland, or place of birth, shared with the ghosts of their ancestors and memories of shared times playing with cousins, siblings and friends.

One of my former students captured this point in an essay they had to present for a course built on the project itself. She said that '[as] the original population and culture of these areas is lost, there is a danger that urban heritage spaces will lose that which makes them attractive and special in the first place. Tourism has changed the very fabric of society in Mediterranean cities where community was everything' (L. Cutugno in Radmilli 2005: 325). This tourism studies graduate highlights the community aspect in Mediterranean urban spaces which are offset by the monumental buildings around which their lives are centred.

Conclusions

Valletta is a UNESCO World Heritage Site that is unique in its physical structure and style. It is a monumental city with an equally imposing past and these grandiose qualities of the city often overshadow the lives of the people who live there and care for it. The various anecdotes and experiences outlined above essentially point in a very particular direction. Most residents – even past residents who had to leave – still have a very strong sense of identity that is clearly attached to the city and their sense of place (real or remembered) that is located within the walls of Valletta. It is breaking their heart to see Valletta become a ghost of its former glamorous past where the streets were alive and populated by the residents with the smells, sights and sounds of everyday life. So there is a drive and desire to get the city back on its feet again. There is also an equal desire on the side of the official political groups to get Valletta going again and this has been witnessed in many ways through the regular events that have been organized in the city over the past few years. There is also a national debate about new plans to redesign and reconstruct the 'city gate' area which is one of the primary access points to the city. It is clear from these ongoing debates that people care about the decisions that will be taken, and the eminent architect Renzo Piano has been called in to help with the design. Following from Bachelard (1994) this swaying from large to small, official to mundane and palace to street is what makes life so rhythmic and musical, although much of the rhythm is being lost as many of the smaller details are being lost.

But what needs to be done to bridge the proverbial gap between these different user groups? And what lessons can policy-makers learn from these experiences? This idea of closer collaboration and actively working with the people on the ground is what Bianchi proposes and this came out clearly in Slavko's experience and is also reflected by the students' critique which seems to call for closer contact with the local residents and the stories that they have to tell about their respective sense of belonging. Sense of place and belonging either has a history or needs to be earned. Dramatic events in the history of Valletta have led to a decline in its population, when ironically it is the community that makes the Mediterranean city and that lies at its heart. The Mediterranean Voices Project was one among others that worked as a multisited, multivocal and multidisciplinary project based on intangible heritage in an urban context and this very same sense of identity was at the heart of our work. Contested spaces and identities are a reality in many parts of the Mediterranean albeit in varying formats and extents and for different political reasons. But many of the situations existent in the cities today are rooted in the past, and a past that is sometimes shared with other places. Likewise some communities are experiencing the same problems that other groups experienced in the past, so by talking about things, and comparing different situations on the ground the project also highlights a lot of commonalities – not all of which are tragic by nature.

Sense of place may be hard to define due to its fluid nature but it certainly exists and people are very attached to it. Memory is often riddled with nostalgia but it can also inform current and future decisions for the enhancing of a community as well as contributing to community spirit in no uncertain terms. Turner tells us that 'as members of society, most of us see only what we expect to see, and what we expect to see is what we are conditioned to see when we have learned the definitions and classifications of our culture' (1967: 95). Researchers have the interesting task of understanding these definitions and working with civil society and policy-makers to render culture (in the broadest sense) accessible without sacrificing authenticity and also respecting change and the dynamics inherent to any community.

Various projects like Med-Voices target different user groups including young students some of whom may well be the policy-makers of the future. Digitization, ICT and multimedia have opened up a wealth of opportunities for the recording, processing, storage and dissemination of information that broaden the scope of these projects dealing with tangible and intangible heritage. The idea of networks also emerges from these projects – not only between the various consortia – but also between official institutions and local stake-holders, ultimately building a bridge between various governing bodies (including the official EU bodies) and the people on the ground who want these initiatives, as well as civil society that often needs to benefit from these initiatives, connections and networks.

I have spoken at length about the personal experiences of a few people involved in the project and film as well as some reactions from one of our main audiences, and one last vignette will help tie it all together. A former student who was actually an ERASMUS exchange student from Sweden attending this same class in Malta stated that 'fragile inter-communal ties that have often been around for generations and the coexistence of different ethnic groups and religions, have periodically degenerated into religious conflicts, racism and violence. *Living together in Mediterranean cities has traditionally been dependent on the capacity to create cultural and social space together*' (J. Smedlsund in Radmilli 2005: 340. Emphasis my own).

We have seen the effects of increasing public dialogue in better informed societies, where public consultations are often de rigeur. While they tend to attract all sorts of armchair critics, tight networks and active engagement between and within different groups could take us towards a more solid approach in managing sense of place and identity relating to all kinds of heritage. These approaches should aim to support a society which can fight for its heritage in sustainable ways that ensure that these forms of heritage are nurtured.

De Certeau echoes the relationship between city and life when he tells us that: 'the childhood experience that determines spatial practices later develops its effects, proliferates, floods private and public space, undoes their readable surfaces, and creates within the planned city a "metaphorical" or mobile city, like the one Kandinsky dreamed of: "a great city built according to all the rules of architecture and then suddenly shaken by a force that defies all calculation"' (1988: 110).

Acknowledgements

This project formed part of the EuroMed Heritage II program which was coordinated by Dr Julie Scott, Dr Raoul Bianchi and Professor Tom Selwyn. I wish to thank my colleague Mark Casha who worked with me on the project and on the original version of this paper delivered at the EAA conference in September 2008. Special mention and thanks goes to the Maltese project coordinators Mr Carmel

Fsadni and Prof Edward Zammit. University of Malta was the institutional partner. Thanks also to the many students who offered their comments, and these include the former Tourism Studies students who participated in the Mediterranean Voices course, the students from the MA in Cultural Heritage Management in Morocco (TEMPUS program) and the students from Bristol University reading for their MA's in Archaeology for Screen Media, Historical Archaeology, Landscape Archaeology, and Twentieth-century Conflict Archaeology, who kindly sent some very interesting feedback. I also wish to thank the film director Edward Said and his team, especially Slavko for a speech that was right on the mark. And thanks to all those who contributed to this project in their own way, and to the other teams on the Mediterranean Voices project. Finally, I would like to thank John Schofield and Rosy Szymanski for the opportunity to participate in the conference session from which this chapter derives.

References

Bachelard, G. 1994. *The Poetics of Space*. USA: Beacon Press.

Bianchi, R. 2005. Euro-Med Heritage: Culture, Capital and Trade Liberalisation – Implications for the Mediterranean City. *Journal of Mediterranean Studies* 15(2), 283–318.

Blok, A. 2001. *Honour and Violence*. London: Polity Press.

De Certeau, M. 1988. *The Practice of Everyday Life*. California: California University Press.

Ilhna Beltin 2005. (Rachel Radmilli and Mark Casha. Film produced for the MedVoices project).

MEPA 2002. http://www.mepa.org.mt/Census/Pyr%20Archive/pyrtab/Valletta. htm [accessed: 3 March 2010].

Radmilli, R. 2005. The Educational Output of EU-funded Projects: Mediterranean Voices Within the Academic Context. *Journal of Mediterranean Studies* 15(2), 319–44.

Scott, J. 2005. Imaging the Mediterranean. *Journal of Mediterranean Studies* 15(2), 219–44.

Selwyn, T. 2005. Mediterranean Counterpoint: Themes and Variations in the Medvoices Database. *Journal of Mediterranean Studies*, 15(2), 245–82.

Selwyn, T. and Radmilli, R. (eds) 2005. *Journal of Mediterranean Studies* 15(2). (Mediterranean Voices special edition).

Szerszynski, B. and Urry, J. 2006. Visuality, Mobility and the Cosmopolitan: Inhabiting the World from Afar. *British Journal of Sociology* 57(1), 113–31.

Turner, V. 1967. *The Forest of Symbols*. USA: Cornell University Press.

Chapter 13

Topophilia, Reliquary and Pilgrimage: Recapturing Place, Memory and Meaning at Britain's Historic Football Grounds

Jason Wood

By way of introduction, let me first reveal the following passage from Nick Hornby's bestseller *Fever Pitch* explaining Hornby's search for a flat in 1989 close to his beloved Arsenal Stadium in Highbury, north London:

> ... living within walking distance of the ground was the fulfilment of a pitiful twenty-year ambition ... It was fun looking. One flat I saw had a roof terrace which overlooked a section of the front of the stadium, and you could see these huge letters, 'RSEN', no more than that but just enough to get the blood pumping. And the place we got gazumped on was on the route that the open-top bus takes when we win something ... In the end we had to settle for somewhere a little less spiritual overlooking Finsbury Park, and even if you stand on a stool and stick your head out of the window you can't see anything ... But still! People park their cars in our road before the game! And on a windy day the tannoy is clearly audible, even from inside the flat, if the windows are open! ... And best of all, just a few days after moving in, I was walking down the road – *this really happened* – and I found, just lying there, filthy dirty and somewhat torn but there nonetheless, a twenty-year-old Peter Marinello bubblegum card. You cannot imagine how happy this made me, to know that I was living in an area so rich in archaeological interest, so steeped in my own past. (Hornby 1992: 210–11)

Hornby's emotive words urge us to consider both the tangible and intangible heritage of football grounds; how they are valued as emblematic of aspiration and achievement; and to understand the intense sense of identity and of place which they convey in popular culture. In attempting to deconstruct or 'excavate' the above passage, this chapter will seek to explore football fans' emotional and subjective attachment to these cherished locations; the different ways in which this attachment is expressed; the importance of such places as repositories for, and conduits of, public memory; and the potency of some former grounds to create new interest in history and heritage, and to generate new tourist markets and destinations.

Place and Memory

> ... sport leaves its imprint on culture, it also leaves its impress on the *landscape*.
> (Bale 1982: 1)

Place plays a significant role in sport and leisure and John Bale, the eminent sports geographer, has developed these links in a number of important books and papers. Bale brought the term 'topophilia', first popularized by Yi-Fu Tuan (1974), into British academic use some years ago as a short hand for that emotional attachment to place, that aura compounded of remembrance of past pleasure and pain that attaches football fans to stadiums (see, for example, Bale 1989). In later works Bale has explored the distinctive character and pervasive form of sports landscapes arguing that football grounds are architectural repositories of history, acting as clear sources of continuity between different generations of fans and players (Bale 1993, 1994). Bale quotes a Chester City fan on the closure of the club's ground in 1990:

> Sealand Road has been part of my life for 30 years; it's more than a football ground; it's a way of life not just to me but to thousands of people alive and dead whose lives have revolved around a match at The Stadium. It is more than bricks and mortar, it's almost something spiritual. (Bale 1993: 65)

Similar sentiments are expressed in a series of quotations contained in Gary James' book *Farewell to Maine Road*. The former Manchester City chairman Eric Alexander, speaking after the ground had staged its last Manchester derby, described how:

> I simply stood there and looked around. I'm not a sentimental man, but I did consider the events that had occurred at the ground. I could picture my father at key moments in his life which, inevitably, all involved City and Maine Road ... I don't believe I had ever stood and thought about the ground and my memories in that way before. Normally, you watch the match or work in the offices and that's it. You don't often stand and think about what you've seen over the years. The great players ... the terrific matches ... There's more to life than a building and a football team, but that doesn't mean they are trivial. For some supporters they are everything. (James 2003: 23)

Steve Heald's familial ties to the ground are tinged with sadness that his children will never feel the unique experience of going to matches the way he did. 'They'll never consider Maine Road as part of their heritage, or birthright, the way I do' (James 2003: 315).

Football grounds, therefore, have the power to stir hearts and minds and to evoke and anchor strong visual and social memories. This relationship is infused with past and present experiences, with sensual memories of how they look, sound, feel, smell and taste. *The Guardian* newspaper journalist and footballing obsessive Harry Pearson, recalling his early years at Ayresome Park in Middlesbrough: 'Ayresome ... was cavernous and cold. The seats were hard, the air filled with the stink of frying onions and ale and the frustrated growl of the crowd, always teetering between frustration and joy' (Gabie and Pearson 2001).

As an archaeologist and heritage consultant, I first approached these themes from a conservation management perspective with an emphasis on historic 'sportscapes'. Bale (1989) introduced the term 'sportscapes' to describe places designed specifically with sport in mind. My interests lie in the history and heritage of such places. In 2001 English Heritage commissioned a pilot study looking at various aspects of the history and heritage of sport in Manchester in advance of the Commonwealth Games held in the city in 2002 (Wood 2005a; see also Inglis 2004). Since then, I have developed research interests in the rapidly developing fields of public history and public understanding of the past. These areas become especially important and challenging when researching former sports places – how to realize the value of what survives; how to appreciate what these survivals mean to people; how to make informed and appropriate choices of what to retain and how to adapt; how best to memorialize those valued things we have lost or will lose; and how to mark and celebrate the tangible and intangible heritage of our sporting past. In short, it means understanding the public representation of sport through its history and heritage, through its material and non-material culture, and promoting programmes that offer a more inclusive reach in terms of community participation (see, for example, Wood 2005b).

Loss and Change

Britain's sporting heritage is a finite and irreplaceable resource, but despite its distinctiveness and authenticity decades of under-appreciation and lack of protection have taken their toll, resulting in loss of, or damage to, some famous and popular landmarks. It is arguable that the heritage sector, in failing to recognize the changing perceptions of historic sports places, has responded inadequately, belatedly and inconsistently. Of course loss and change have occurred throughout history. Sports venues of all kinds have been subjected to redevelopment dictated by the needs of clubs, players and spectators. But more often than not these developments were relatively modest and undertaken with a sense of proportion, and a *sense of place*. Today, the scale of loss and change is unprecedented. Controversial closure and disposal of historic sports places by public and private bodies continues unabashed – to raise revenue, reduce expenditure, pursue specific and transitory visions of current 'best practice' as promoted by developers, or comply with health and safety standards – and with too little regard for their historic and heritage values. This has led to increased planning casework, political interest or interference and media representation (often misrepresentation) but also to a growing number of public protests and demonstrations (Walton and Wood 2007).

Nowhere is the impact greater, or felt more acutely, than in the sport of football. The recommendations by Lord Justice Taylor (1990) for all-seater stadiums following the Hillsborough disaster in 1989 spawned a concentrated and comprehensive period of demolition, redevelopment and relocation of British football grounds. Many former grounds have now disappeared without trace below housing estates, retail parks and supermarkets (Figure 13.1) while others lie vacant or partially demolished and overgrown (Figures 13.2 and 13.3). Despite improvements in design terms the poor quality, setting and location of some of the new facilities that replaced them has led some to echo sociologist George Ritzer's thesis (Ritzer 1993) and criticize the so-called 'McDonaldization' of sports buildings. As one complainant to *The Observer* newspaper put it, soon 'every city will have the same bland and characterless sports arena and to hell with heritage

Figure 13.1   **Sunderland's Roker Park hosted its last game in 1997 before the club moved to the Stadium of Light. The site now lies buried below a housing estate centred on a small public play area where ironically 'NO FOOTBALL OR OTHER BALL GAMES (ARE) ALLOWED'. New street names commemorate the former stadium, but note the absence of a 'Relegation Close'**

Source: Photograph: Jason Wood

and history'. Contributors to football fanzines bemoan the soulless 'container architecture' of their 'identikit' stadiums. Gone is the oddity of the shapes of the different grandstands that gave football grounds that sense of special location and provided each ground with its own characteristic identity.

Figure 13.2   Stoke City's Victoria Ground also hosted its last game in 1997 before the club moved to the Britannia Stadium. The grandstands were soon demolished and the site now lies abandoned, awaiting redevelopment. Interestingly, for months after the closure, flowers would be laid on the former centre spot at 3pm on match days (John Bale, personal communication)

Source: Photograph: Jason Wood

Figure 13.3   **The decaying hulk of 'The Vetch' – the home of Swansea
City until 2005. The blocked turnstiles, barbed wire and
surveillance cameras could not be less welcoming than the
prison next door. Denied access to the pitch, local kids have
to satisfy themselves with a kick-about in the streets outside**

Source: Photograph: Jason Wood

Reliquary

As is well known, football fans can symbolize the love of their team or club through a wide range of material culture. Many hoard old programmes, scorecards, ticket stubs, photographs and scarves that signify their particular team or club's history and their personal history of attachment to it. So it is perhaps unsurprising that when the club stadium is redeveloped or relocated, a seat, a sign and even pieces of concrete and strips of turf also become treasured possessions. The collection of such relics is legitimized by the growing number of football museums and fuelled generally by the global market in sports memorabilia and ephemera. Symbolic references to an idealized past they may be, but is this 'tribal bric-à-brac', as the anthropologist Desmond Morris has described it (Morris 1981: 284); are these spoils of sport – this reliquary – heritage or 'heritat'? What is clear is that these fragmented relics yield powerful souvenirs to their owners and other fans and reveal a strong desire for a tangible and authentic past.

In the 1990s as the post-Taylor redevelopment or closure of football grounds began to accelerate, so the souvenir hunters moved in, sometimes, as at Airdrie's Broomfield Park, before the last game was over! In more recent years, clubs have taken more precautions to safeguard important items of memorabilia and material and begun to organize auctions in an attempt to regulate the souvenir hunters and realize the monetary value of such memorabilia for themselves.

The biggest of these auctions was in July 2006 when Arsenal bid farewell to Nick Hornby's beloved Highbury, the club's home since 1913. The auctioneer stood on a covered stage on the pitch surrounded by accoutrements more usually associated with a rock concert – a big screen and loud speakers, with the stage itself a tangle of wires and computer terminals as bids were received online from all over the world (Figure 13.4). The much-hyped red plastic wheelie bins with Arsenal insignia were sure to find a home: purchase these, the auction catalogue said, and 'be the envy of your neighbours!' Described by the auctioneer as 'la crème de la crème of the auction' and 'guaranteed to get you an interview on the BBC' the wheelie bins sold for £600 each. Signs proved a popular but expensive investment, especially those for 'GENTS TOILETS' and 'NO ALCOHOL BEYOND THIS POINT'. Tiles from the toilets sold for over £200 each: there were plenty still on the walls but I resisted the temptation to bypass the auction with a chisel. I asked one woman, who did buy a tile, what she intended to do with it. She told me she would 'put it on the mantelpiece where all my important things go along with family photographs, and look at it every day' (Wood 2006).

Figure 13.4 **The stage is set for the Highbury auction. Almost 3,000 people attended, with about half as many again bidding via the Internet against those in the stadium, peering down from the upper tier of the east stand**

Source: Photograph: Jason Wood

Marking and Celebrating Former Grounds

Commemorative creations to former football stars and past events are on the increase, but are we equally alive to the potential of marking and celebrating former football grounds?

In the Manchester study (above), over 70% of people interviewed thought it was important to commemorate those places where historic sporting events happened – for the future, for the community and for children in particular. People offered many ideas about how sporting memories could be kept alive, including celebration of customs, traditions, routines and practices that they associated with such places. However, the active promotion of forward-looking strategies that are sensitive to the richness of sports history and its personalities, and investment in 'live' schemes and events that reanimate places and promote contemporary sport, were both favoured strongly (Wood 2005a: 142). This was reinforced during a seminar held in Oldham to discuss ways in which sport and culture can create a sense of local pride and belonging. In the resulting report 'celebration' is described as 'the x-factor': 'Community celebration, such as street festivals and events organized by museums, galleries, sports stadiums and so on, is a brilliant tool for community engagement simply because people have fun' (DCMS 2004).

To date the marking of former football grounds has been expressed in a number of subtle and celebratory ways. There is a tradition of street names (Figure 13.1) and pub names being used to commemorate the former presence of a club or stadium. Historic plaques are becoming more common, as is the appropriation of relics from old stadiums to new. Even more interesting is the creation of evocative public artworks associated with ground relocations. Moreover, such associations can have significance out of all proportion to the character of the surviving buildings, if any remain at all.

The retention by clubs of relics from old stadiums for use or ornament in new structures is a growing trend. This kind of adaptive reuse sees turnstiles given a new lease of life and boardroom panelling cannibalized and reset into new surroundings. For example, in Huddersfield, over 40 Leeds Road turnstiles were recycled in the building of the Alfred McAlpine Stadium (now the Galpharm Stadium); parts of the latticework balcony from the stands at Roker Park in Sunderland are now displayed in the car park at the Stadium of Light; while the embossed headstone reading 'Rovers FC' from a turnstile block at Ewood Park in Blackburn is now cleaned and rebuilt into a wall behind a statue of Jack Walker, the club's modern benefactor.

One of the most fascinating commemorative art projects is *The Trophy Room* on the redeveloped site of Ayresome Park, the historic home of Middlesbrough Football Club (Gabie and Pearson 2001). When the club moved out of the stadium in 1995, there was room for over 100 houses to be built where the pitch and stands had been. The project, sponsored by Cleveland Arts and funded by Wimpey Homes and Middlesbrough Council, involved the incorporation of artworks by Neville Gabie into the new housing estate as symbolic allusions to the former ground.

**Figure 13.5 The Trophy Room – a pair of bronze football boots on the front
doorstep of 18 The Holgate mark the former Ayresome Park
centre spot**

Source: Photograph: Jason Wood

So, one of the penalty spots is marked by a bronze football in a resident's front garden; stainless steel studs set into the tarmac of driveways mark the touchlines and centre circle; a pair of bronze football boots mark the centre spot (Figure 13.5); a bronze scarf and bronze jumper mark two of the corner flags; the words 'AWAY' and 'ENCLOSURE' are sandblasted into brick boundary walls; there is even a reminder in the form of a bronze pitch puddle covered in stud marks from where Pak Do Ik struck his match-winning goal for North Korea in the victory over Italy in the 1966 World Cup.

The accompanying pamphlet contains photographs of these installations juxtaposed with historic photographs of the relevant parts of the former stadium. Sometimes the photograph becomes the artwork itself as in the case of the images taken within residents' houses. The cover is illustrated with flakes of red paint from the ground, photographed against a white background – the team colours. The inside cover comprises a plan of the former ground overlain by a plan of the housing estate. Harry Pearson provides an amusing text where we are immediately drawn into familial reminiscences of Grandad crawling under the turnstiles as a boy and later a young Harry being lifted over them: a familiar trans-generational attachment and 'rite of passage' for most football fans entering 'the world of men' (Gabie and Pearson 2001). We are also treated to a resident's view. Robert Nichols, the first to buy one of the houses at Ayresome and editor of the fanzine *Fly Me to the Moon*, writes:

> I'll never forget the last day at Ayresome Park. At the final whistle on the final league match, celebration at near promotion was tempered by a lump in the throat sadness. Over twenty thousand pairs of eyes scanned the familiar surroundings one last time. The temple of our dreams would soon be shattered as surely as those hopes and aspirations had so often been in the past. It was all history now. Not for me however. I'd decided from the very start that this place meant far too much to let go. And no the offer of a chunk of turf or an old seat wrenched from a stand just wasn't enough. This was the place where I had come to worship for years, it was the very heart of our town. It was almost sanctified ground. This was the place I had to live. (Gabie and Pearson 2001)

The project reveals some interesting relationships between past and present and also between artist and audience. Not only did the residents willingly agree to the project; they helped select the artist. *The Trophy Room* is a good example of the ways in which a project can be developed to mark the memories, the events and the home of a local team. On the face of it, the housing scheme is a pleasant but otherwise unremarkable development on an infill brownfield site, but the association with Middlesbrough Football Club, and the public art project that celebrates that association, gives the development a strong sense of place and community cohesion (Wood and Gabie forthcoming). It stands for 'the fans part in what made Ayresome special.' If you go there, and many fans do, the residents will happily engage you in conversation and proudly show off 'their' bits of the 'stadium'.

Pilgrimage and Nostalgia Tourism

This last example is a reminder of the growing realization of the value of sports history and heritage for tourism. Visits to sports venues now make up many tourists' itineraries (see, for example, John 2002); so much so that for the first time the Automobile Association has published a sports atlas of Britain and Ireland locating over 750 venues (Automobile Association 2005). Again, the phenomenon is especially acute at football grounds where levels of public interest and visitation have been motivated particularly in the last decade by the rise and popularity of club museums, halls of fame, stadium tours and even fantasy camps (themed vacations with sporting professionals). Clubs have not been slow in recognizing the economic and commercial benefits of such tourism. For example, with 250,000 visitors a year, Manchester United's museum at Old Trafford is in the same league as English Heritage's top visitor attractions, such as Dover Castle and Osborne House (Wood 2005a: 142). And visits to famous sports venues are not just confined to Britain (see Gammon 2004). For instance, in Barcelona the queue for the tour of the Nou Camp stadium is often much longer than that for Gaudi's Sagrada Familia.

Pilgrimages to historic football grounds are also on the increase (Gammon 2002) with fans gripped by what the writer and stadium expert Simon Inglis has coined 'stadiumitis' – that 'heart-fluttering sense of anticipation' and 'heightened sense of elation' that grips fans as the floodlights and stands loom into view (Inglis 1996: 8). The most chronic sufferers of 'stadiumitis' must be members of the 92 Club, the so-called ground-hoppers whose aim is to attend football matches at the grounds of all 92 professional league clubs in England and Wales. The Club is celebrated on several websites (see, for example, http://www.footballgroundguide.com) where members are invited to share their achievement and experiences and post photographs of themselves at their final ground. And people even travel from elsewhere in Europe to do this. Han van Eijden from the Netherlands proudly completed the 92 and once saw three matches at different grounds on the same day!

The examples of Manchester United, Barcelona and the 92 Club are indicative of the potency of football grounds to attract many visitors from long distances. Surviving historic grounds, and the sites of former stadiums, have the potential therefore to recapture place, memory and meaning, to create new interest in history and heritage and to generate new tourist markets and destinations. However, to date, such developments have largely been ad hoc and the benefits of so-called 'nostalgia sport tourism' and community participation in destination development (such as at Ayresome Park) have been overlooked and undervalued.

**Figure 13.6 Highbury Square nearing completion in February 2009 –
'a quality residential development that reflects the club's
heritage and allows the spirit of Highbury to live on' –
Arsène Wenger**

Source: Photograph: Jason Wood

Let us return again to Arsenal's former ground at Highbury. In September 2009 the last of more than 700 luxury apartments in Highbury Square was completed (Figure 13.6). The shells of the art deco east and west stands survive and have been sympathetically converted, with the famous marble halls and grand staircase in the listed east stand being retained as the entrance to the most exclusive apartments. The pitch has become a communal garden with illuminated water features (see http://www.highburysquare.com). Simon Inglis has asserted that Highbury was 'quite simply … the most balanced and orderly ground in the country. There is not a line out of place; all is in total harmony' (Inglis 1983: 221). The redevelopment scheme is certainly in keeping with this spirit in uniquely seeking to preserve the stadium's historic fabric, footprint and sense of enclosure, as well as capturing the aura and memories of the place, while once again making it a beacon for fresh activity and giving it back a purpose (Wood 2006). In the words of the Arsenal chairman, Peter Hill-Wood: 'Our wish for the development was always to retain more than a passing resemblance to Highbury Stadium and to respect its class and heritage … Although Highbury as a football stadium is now gone, Highbury Square has ensured that our old home will never be forgotten' (http://www.arsenal.com).

My only criticism of the scheme is that I believe the club has underestimated both the future level of public response to the redevelopment and the commercial potential of the growing interest in nostalgia sport tourism. When I asked Arsenal fans about the redevelopment many said they intended to visit the old stadium, especially on match days, 'to pay homage' or 'show their children where Dad used to sit' (Wood 2006). Highbury has great significance for football fans, not just Arsenal fans, and the place is likely to be a huge draw, as the new Emirates Stadium is only two streets away; somewhere, as Peter Hill-Wood, writing in the final Highbury match-day programme, was keen to stress – 'so close you can still take the same train, eat in the same restaurant or go to the same pub and I think that's important for the soul of the football club'. What is also important, I would argue, is that public access to its former home be encouraged and marked and celebrated in some way. At the moment the only public access is a through route between the southern ends of the historic stands during daylight hours. The gardens are gated, although limited access is possible to a memorial garden at the south-east corner. The east stand features a viewing gallery overlooking the gardens but access to this is only available to residents as a hired private function room. No provision has been made for ordinary fans.

Conclusion

Twenty years ago, three members of the University of Surrey's Department of Psychology published a study of the environmental psychology of football grounds, concentrating on the spectator's experience of football and on the places where it was played (Canter, Comber and Uzzell 1989). The survey took place before the major redevelopments at football grounds that began in the 1990s and before the growth of club museums and attendant stadium tours, so questions focused on physical conditions at the grounds and possible improvements, rather than concerns over loss and change or tourism opportunities. Nevertheless, the observations are still revealing and relevant today:

> ... people go to football matches to watch football and ... what really matters [to them] is the standard of play on the pitch, the team and how it is managed. ... While this may be so, these are not the only aspects to play a role in defining the identity of the club. It is also true to say that both the intangibles, such as club history and traditions and the 'feel' of a club and the physical reality of the ground ... make a major contribution to the enduring qualities of the club. (Canter, Comber and Uzzell 1989: 64)

> One of the most important physical symbols of a club is the ground itself. The social history of the club is embedded in and complemented by the grounds. (Canter, Comber and Uzzell 1989: 82)

This sense of belonging, often bound up with family history, is all part of traditional fandom as exemplified in Nick Hornby's *Fever Pitch*, where Hornby's experience of what it is to be an Arsenal fan and of actually being at Highbury are considered just as important as events on the pitch.

In their conclusions, Canter and his fellow investigators put forward six major areas of activity which comprised a positive strategy for dealing with the then crisis in British football. One involved making the public in general and the football supporter in particular more aware of their football heritage:

> The history and heritage of football is an extremely valuable asset which can be used to restore pride, create a sense of identity and belonging, enlist public support and enhance the spectator's appreciation and enjoyment of the sport. (Canter, Comber and Uzzell 1989: 163)

To which we can now add – to instigate tourism. Tourism offers a potent and exciting way of promoting and inspiring regeneration at Britain's historic football grounds, contributing to the place-shaping agenda and combining renewal and innovation with an appeal to tradition and identity. However, to be successful, we need first to raise the benchmark for heritage management of these distinctive environments by finding new ways to protect and enhance them and by mobilizing people's affection for their rediscovery, nostalgia and authenticity. In this respect proper mapping and characterization will be essential to ensure that their value and significance – their very essence – permeates through to generate effective policies so that planning, development and tourism decisions are based on informed knowledge, understanding and respect for what has gone before and people's interest in and attachment to it (Walton and Wood 2007).

We also need to address the current lack of research and coordination in this area. Academic debate is beginning to emerge with publication of a collection of essays entitled *Heritage, Sport and Tourism: Sporting Pasts – Tourist Futures* (Gammon and Ramshaw 2007) which provides the first comprehensive resource on sports heritage as a tourist attraction. What is also required is research to reveal the synergies between history, heritage, sport and tourism through the lens of public representation. In this respect a welcome start has now been made by means of a series of seminars organized by De Montfort University's International Centre for Sports History and Culture, the National Football Museum and Heritage Consultancy Services, and funded by an Arts and Humanities Research Council Research Networks and Workshops Award. A book resulting from the seminars will enable strategic markers to be laid down in this emerging field (Hill, Moore and Wood forthcoming). Further research will involve the convergence of many different and diverse fields and an interdisciplinary approach employing the skills and techniques of psychologists, sociologists, geographers and historians, as well as those working in the historic environment, museum and tourism sectors. In seeking to open out debate in this context and further stimulate the overlap between these various disciplines, key areas for exploration will need to focus on such questions as spirit of place; loss and change; memory and meaning; authenticity and nostalgia; and regeneration and sustainability.

This chapter has highlighted the significance of people's interactions with Britain's historic football grounds. The tradition of football is only one dimension of the historic environment but it is a keenly distinctive aspect of 'place' for spectators, supporters and participants. For those engaged actively, the history and heritage of a football ground is defined as much by the longevity and continuity of use as it is by any physical resources or buildings. In other words, the tradition of football in a place appears to be an important element of what is valued, and this persists long after the venue itself may have gone or any historic structures have been removed by new development. Football, history and heritage interests can be made to work together. Whether it is through improved integration of management and conservation action, or by finding ways to mark and celebrate the tradition of football grounds with their communities (or ideally both), it ought, with imagination, effort and investment, to be achievable as well as beneficial to recapture football's historic sense of place.

References

Automobile Association 2005. *Sports Atlas: Britain and Ireland 2006*. Basingstoke: AA Publishing.

Bale, J. 1982. *Sport and Place: A Geography of Sport in England, Scotland and Wales*. London: C. Hurst and Co.

Bale, J. 1989. *Sports Geography*. London: E. and F.N. Spon [2nd edn, 2003. London: Routledge].

Bale, J. 1993. *Sport, Space and the City.* London: Routledge [reprinted, 2001. Caldwell, New Jersey: The Blackburn Press].

Bale, J. 1994. *Landscapes of Modern Sport.* Leicester: Leicester University Press.

Canter, D., Comber, M. and Uzzell, D.L. 1989. *Football in its Place: An Environmental Psychology of Football Grounds.* London: Routledge.

DCMS 2004. *Bringing Communities Together Through Sport and Culture: Oldham 2004.* London: Department for Culture Media and Sport.

Gabie, N. and Pearson, H. 2001. *The Trophy Room: Ayresome Park.* Middlesbrough: Cleveland Arts.

Gammon, S. 2002. Fantasy, Nostalgia and the Pursuit of What Never Was. In S. Gammon and J. Kurtzman (eds), *Sport Tourism: Principles and Practice.* Eastbourne: Leisure Studies Association, 61–71.

Gammon, S. 2004. Secular Pilgrimage and Sport Tourism. In B.W. Ritchie and D. Adair (eds), *Sport Tourism. Interrelationships, Impacts and Issues.* Clevedon, Buffalo and Toronto: Channel View Publications, 30–45.

Gammon, S. and Ramshaw, G. (eds) 2007. *Heritage, Sport and Tourism: Sporting Pasts – Tourist Futures.* London: Routledge.

Hill, J., Moore, K. and Wood, J. (eds) forthcoming. *Sport, History and Heritage: An Investigation into the Public Representation of Sport.* Woodbridge: Boydell and Brewer.

Hornby, N. 1992. *Fever Pitch.* London: Victor Gollancz [reprinted, 2000. London: Penguin Books].

Inglis, S. 1983. *The Football Grounds of England and Wales.* London: Willow Books.

Inglis, S. 1996. *Football Grounds of Britain.* London: Collins Willow.

Inglis, S. 2004. *Played in Manchester: The Architectural Heritage of a City at Play.* London: English Heritage.

James, G. 2003. *Farewell to Maine Road.* Leicester: Polar Publishing.

John, G. 2002. Stadia and Tourism. In S. Gammon and J. Kurtzman (eds), *Sport Tourism: Principles and Practice.* Eastbourne: Leisure Studies Association, 53–9.

Morris, D. 1981. *The Soccer Tribe.* London: Jonathan Cape.

Ritzer, G. 1993. *The McDonaldization of Society: An Investigation into the Changing Character of Contemporary Social Life.* [4th edn, 2004. California: Pine Forge Press].

Taylor, The Rt. Hon. Lord Justice P. 1990. *The Hillsborough Stadium Disaster 15 April 1989. Final Report.* London: HMSO.

Tuan, Y.-F. 1974. *Topophilia: A Study of Environmental Perception, Attitudes, and Values.* Englewood Cliffs, New Jersey: Prentice-Hall.

Walton, J.K. and Wood, J. 2007. History, Heritage and Regeneration of the Recent Past: The British Context. In N. Silberman and C. Liuzza (eds), *Interpreting the Past V, Part 1. The Future of Heritage: Changing Visions, Attitudes and Contexts in the 21st Century.* Brussels: Province of East-Flanders, Flemish

Heritage Institute and Ename Center for Public Archaeology and Heritage Presentation, 99–110.

Wood, J. 2005a. Talking Sport or Talking Balls? Realising the Value of Sports Heritage. In D. Gwyn and M. Palmer (eds), *Understanding the Workplace: A Research Framework for Industrial Archaeology in Britain*. Leeds: Association for Industrial Archaeology, 137–44. [Special Issue of *Industrial Archaeology Review* 27 (1)].

Wood, J. 2005b. Olympic Opportunity: Realizing the Value of Sports Heritage for Tourism in the UK. *Journal of Sport Tourism* 10 (4), 307–21 [reprinted, 2007, in S. Gammon and G. Ramshaw (eds), *Heritage, Sport and Tourism: Sporting Pasts – Tourist Futures*. London: Routledge, 87–101].

Wood, J. 2006. Highbury Auction: The Final Curtain Call for Fans of Arsenal FC. [Online]. Available at http://www.culture24.org.uk/nwh_gfx_en/ART39218.html [accessed: 21 November 2009].

Wood, J. and Gabie, N. forthcoming. The Football Ground and Visual Culture: Recapturing Place, Memory and Meaning at Ayresome Park. *International Journal of the History of Sport*.

Index

Page numbers in italic (eg. *152*) refer to a figure, a map or a table.